THREATENING DYSTOPIAS

A Volume in the Series
Cornell Series on Land: New Perspectives in Territory, Development, and Environment
Edited by Wendy Wolford, Nancy Peluso, and Michael Goldman

A list of titles in this series is available at cornellpress.cornell.edu.

THREATENING DYSTOPIAS

The Global Politics of Climate
Change Adaptation in Bangladesh

Kasia Paprocki

CORNELL UNIVERSITY PRESS   ITHACA AND LONDON

Copyright © 2021 by Cornell University

All rights reserved. Except for brief quotations in a review, this book, or parts thereof, must not be reproduced in any form without permission in writing from the publisher. For information, address Cornell University Press, Sage House, 512 East State Street, Ithaca, New York 14850. Visit our website at cornellpress.cornell.edu.

First published 2021 by Cornell University Press

Library of Congress Cataloging-in-Publication Data

Names: Paprocki, Kasia, author.
Title: Threatening dystopias: the global politics of climate change adaptation in Bangladesh / Kasia Paprocki.
Description: Ithaca [New York] : Cornell University Press, 2021. | Series: Cornell series on land: new perspectives in territory, development, and environment | Includes bibliographical references and index.
Identifiers: LCCN 2021011479 (print) | LCCN 2021011480 (ebook) | ISBN 9781501759154 (hardcover) | ISBN 9781501759161 (paperback) | ISBN 9781501759178 (pdf) | ISBN 9781501759185 (epub)
Subjects: LCSH: Climatic changes—Political aspects—Bangladesh. | Climatic changes—Economic aspects—Bangladesh. | Climatic changes—Social aspects—Bangladesh. | Climatic changes—Political aspects. | Climatic changes—Economic aspects. | Climatic changes—Social aspects.
Classification: LCC QC903.2.B3 P37 2021 (print) | LCC QC903.2.B3 (ebook) | DDC 363.738/74561095492—dc23
LC record available at https://lccn.loc.gov/2021011479
LC ebook record available at https://lccn.loc.gov/2021011480

To the memories of Karunamoyee Sardar and Xulhaz Mannan,
and to the brave activists who honor them by continuing their struggles

Contents

Acknowledgments ix
List of Acronyms xiii

Introduction 1

1. "Sluttish, Careless, Rotting Abundance": Prehistories of a Climate Dystopia 23
2. Threatening Dystopias: Development and Adaptation Regimes 52
3. Opportunity/Crisis: Knowledge Production and the Politics of Uncertainty 78
4. The Social Life of Climate Science: Circulations of Knowledge and Uncertainty in Development Practice 98
5. Autopsy of a Village: Agrarian Change after the Shrimp Boom 118
6. "We Have Come This Far—We Cannot Retreat": Adaptation, Resistance, and Competing Visions of Transformed Futures 158

Conclusion: Climate Justice and the Politics of Possibility 190

Methodological Appendix 199
Glossary 205
Notes 207
Bibliography 219
Index 245

Acknowledgments

My first thanks go to the members and organizers of Nijera Kori, to whom I owe the greatest debt for their boundless generosity in teaching me about development, agrarian change, social mobilization, and their own day-to-day struggles over the power to shape each of these. I first visited Khulna in 2012 at the urging of Nijera Kori organizers. I thank them for leading me to this project and to my initial questions. I have used pseudonyms to refer to Nijera Kori members throughout the book, and thus, unfortunately, I am unable to thank them here by name. I feel a tremendous debt to several individuals personally as well as the movement as a collective. They welcomed me into their communities and their lives, gave me their time and ample fresh green coconuts, and taught me how to plant rice—(and were good-natured about the sloppy results). I hope that I have honored these members and the Nijera Kori staff in Paikgachha, Dumuria, Khulna, and Dhaka and the experiences they have shared with me in my attempts to represent them.

Among these activists, Rezanur Rahman Rose has been a treasured research collaborator, colleague, and friend. I have been honored to have the opportunity to observe his work as a community organizer in villages all over rural Bangladesh, from large crowds to intimate conversations. His incisive analysis, commitment to authentic understanding, and passion for the movement have deeply inspired me. I feel incredibly privileged to have had the opportunity for over a decade to observe closely Khushi Kabir's unparalleled leadership, determination, and perseverance in fighting for the civil and human rights of women, laborers, and rural communities. Her strength and political will have made me aspire me at every step to understand better and to work harder. I will be forever grateful to both Rose and Khushi for inviting me into the movement and into their families.

This work was supported by the National Science Foundation, the Social Science Research Council, the Fulbright-Hays Program, the American Institute of Bangladesh Studies, Inter Pares, the Department of Geography and Environment at the London School of Economics (LSE), and the following programs at Cornell: the Atkinson Center for a Sustainable Future, the Mario Einaudi Center, the Department of Development Sociology, and the South Asia Program. I received vital support and knowledge from several librarians and archivists at the National Library, Kolkata; the National Archives of India in Delhi; the Center for Environmental and Geographic Information Services (CEGIS) in Dhaka; the library at the Institute of Water and Flood Management at Bangladesh University of Engineering and Technology in Dhaka; the British Library; the Wageningen University Library; and the US National Archives at College Park, MD. At the Wageningen library, Zach Lamb was a terrific research collaborator and interlocutor. Additionally, Gertrud and Helmut Denzau shared with me their insights and their impressive personal archive of materials related to the Sundarban region, which is housed in their home in northern Germany.

Some of the arguments and ethnographic material regarding the adaptation regime were previously published in the Annals of the American Association of Geographers. I would like to thank the anonymous reviewers for that publication as well as James McCarthy for his expert editorial guidance in that early iteration of the work. This work was strengthened through valuable engagement with colleagues at several workshops and seminars, including the SSRC InterAsian Connections Frontier Assemblages Workshop, the Denaturalizing Climate Change: Perspectives for Critical Adaptation Research workshop, the School of Geography and the Environment at Oxford University, the Institute of Development Studies, the Department of Geography at King's College London, the University of Liberal Arts Bangladesh, the Bengal Institute for Architecture, Landscapes and Settlements, the ESRC STEPS Centre Symposium on the Politics of Uncertainty, and the Political Ecology Seminar series at the University of Bristol.

Friends and colleagues at Cornell offered intellectual challenges and support, in particular Andrew Amstutz, Ian Bailey, Sara Keene, Divya Sharma, and Greg Thaler, as well as the members of Wendy Wolford's lab group. I greatly benefited from discussions with several Cornell faculty members throughout my time there, including Shelley Feldman, Fouad Makki, Chuck Geisler, Marina Welker, Ray Craib, and Anne Blackburn. Durba Ghosh and Phil McMichael expanded this work through their thoughtful and attentive reading and helping me to see the broader intellectual stakes of the project. Since we met in 2006, Jason Cons has worked alongside me in Khulna and offered incomparable advice, mentorship, and friendship. He, Erin Lentz, and Mira Cons have been family through the most difficult and the most joyful times. It has been my greatest fortune to have the opportunity to learn from Wendy Wolford. Her scholarship has inspired many of my most urgent intellectual and political questions and commitments. She has modeled and mentored me in pursuing the best of what it means to be a feminist in the academy. I aspire to the level of commitment, intellect, and passion that she embodies in her role as an educator, mentor, scholar, and activist.

At the LSE, I have been lucky to find a community of colleagues who have been generous with insight as well as a joy to work with. This manuscript has particularly benefited from discussions with Megan Black, Ryan Centner, Tim Forsyth, Kathy Hochstetler, Naila Kabeer, David Lewis, Claire Mercer, Austin Zeiderman, and participants in the Social Life of Climate Change seminar series.

ACKNOWLEDGMENTS

Several other institutions have provided intellectual communities for me during the process of research and writing. I am particularly thankful for enriching affiliations with the Global Change Program at Jadavpur University in Kolkata (especially Joyashree Roy), the International Centre for Climate Change and Development in Dhaka (especially Saleemul Huq), and the Institute for Public Knowledge at New York University. Yale's Program in Agrarian Studies and its Environmental Anthropology Collective provided an invaluable intellectual community, and I am particularly grateful to Shivi for inviting me to participate in both. I have greatly benefited from discussion, feedback, and provocations from Shapan Adnan, Majed Akhter, Nikhil Anand, Zach Anderson, Hillary Angelo, Ben Belton, Debjani Bhattacharyya, Jun Borras, Jim Boyce, George Caffentzis, Alejandro Camargo, Nusrat Chowdhury, Anjan Datta, Michael Eilenberg, Sylvia Federici, David Gilmartin, Kian Goh, Jesse Goldstein, Meghna Guhathakurta, Shubhra Gururani, Tariq Jazeel, Malav Kanuga, Naveeda Khan, Sarah Knuth, Alex Loftus, David Ludden, Anu Muhammad, Dilshanie Perera, Shaina Potts, Anne Rademacher, Dina Siddiqi, and Michael Watts. Hugh Brammer was generous with his maps and knowledge of the region. Emily Barbour, K. Sivaramakrishnan, and Chris Small have each provided invaluable feedback on chapters of the book. Anjan Datta, Naila Kabeer, Khushi Kabir, David Lewis, and Claire Mercer have each read the manuscript in its entirety and provided crucial feedback for revision. Nancy Peluso edited the book on behalf of the Cornell Press Series on Land, reading multiple drafts and sharing hours of phone calls between London and Jakarta. Her insights have improved the final manuscript tremendously, along with those of the two anonymous reviewers solicited by the publisher.

During fieldwork, many friends and colleagues helped make Bangladesh home: Sara Afreen, Nasrin Akter, Manosh Chowdhury, Taheera Haq, Dina and Hameeda Hossain, Khushi Kabir, Rohini Kamal, Mahrukh Mohiuddin, Helen Naznin, Rose Rahman, Kamar Ahmed Saimon, Shateel Bin Salah, Corey Watlington, Scott Wenger, Shamyu, Tuhin, and Tushar. Xulhaz Mannan was a great friend who supported my work in Bangladesh with advice, support, and companionship for over ten years. He introduced me to the things that made me fall in love with Dhaka, and he deepened my understanding of the cultural politics of Bangladesh and the challenges faced by the country's sexual minorities. I wish I could have told him how much he would be missed.

I have endless thanks and admiration for Liz Koslov and Rebecca Elliott and can't believe my luck in finding each of these brilliant women at critical moments in my career. Their scholarship inspires me, and their friendship and support sustain me. The majority of this book was written while sitting across from one or the other of them at coffee shops and dining room tables in New York City and London. They, along with Daniel Aldana Cohen, have made up the best writing group I could have ever hoped for. Together, these three have deepened my appreciation of the politics of climate justice and pushed me to think in new ways about the broader stakes of my work.

Finally, I have the deepest gratitude for my extended family for love, support, and good humor throughout the process of writing this book. Thank you to Anna Marschalk-Burns, Steve Paprocki, Kate Wesley, Ann Haugejorde, Michael and Marais Bjornberg, Peter Schmidt, and the entire Paprocki family. Rebecca Elliott and Andrew and Ruth Duncan have made London home and have filled our life here with joy and family. Anders Bjornberg has been my partner not only in life, but also in every aspect of the work that is represented here since our first visit to Bangladesh in 2006. It would be impossible to list all the ways in which he contributed to this book as a reader, interlocutor, research assistant, editor, and companion. On the several occasions he has visited Khulna with me, his affable smile, sensitive questions, and charming ease with the Bengali language helped endear me to the residents there, many of whom affectionately refer to him as *Dula Bhai* (Brother-in-Law). The manuscript has been immeasurably strengthened through his ethnographic intuition, intellectual perceptiveness, thoughtful political instincts, and brilliant creative imagination. I am infinitely thankful to him for filling my work and my life with laughter, adventure, and love.

Acronyms

BRAC	Formerly known as the Bangladesh Rural Advancement Committee
BRRI	Bangladesh Rice Research Institute
BWDB	Bangladesh Water Development Board
CBOT	Community-Based Oral Testimony
CEP	Coastal Embankment Project
CGIAR	Formerly known as the Consultative Group for International Agricultural Research
COP	Conference of Parties (United Nations Climate Change Conference)
CSA	Climate Smart Agriculture
DDP	Delta Development Project
DFID	United Kingdom's Department for International Development
DoAE	Department of Agricultural Extension
EPWAPDA	East Pakistan Water and Power Development Authority
FAO	Food and Agriculture Organization of the United Nations
FAP	Flood Action Plan
GEF	Global Environment Facility
IPCC	Intergovernmental Panel on Climate Change
IWFM	Institute of Water and Flood Management at the Bangladesh University of Engineering and Technology
KJDRP	Khulna-Jessore Drainage Rehabilitation Project
LGED	Local Government Engineering Division
MoEF	Ministry of Environment and Forests
RTI	Right to Information
TRM	Tidal River Management
UNDP	United Nations Development Programme
UNFCCC	United Nations Framework Convention on Climate Change
USAID	United States Agency for International Development
WARPO	Water Resources Planning Organization

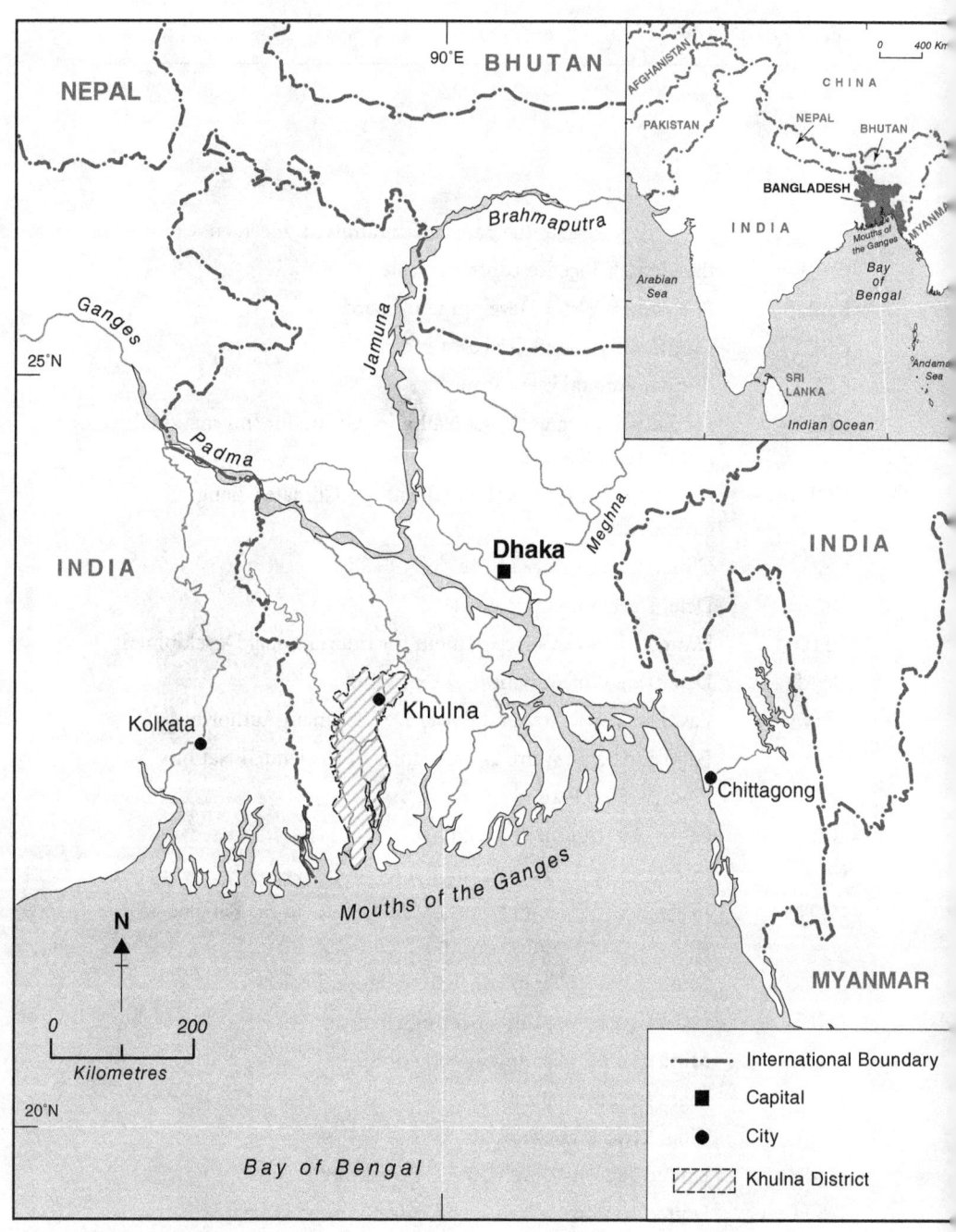

MAP 1. Map of Bangladesh.

Created by Mina Moshkeri.

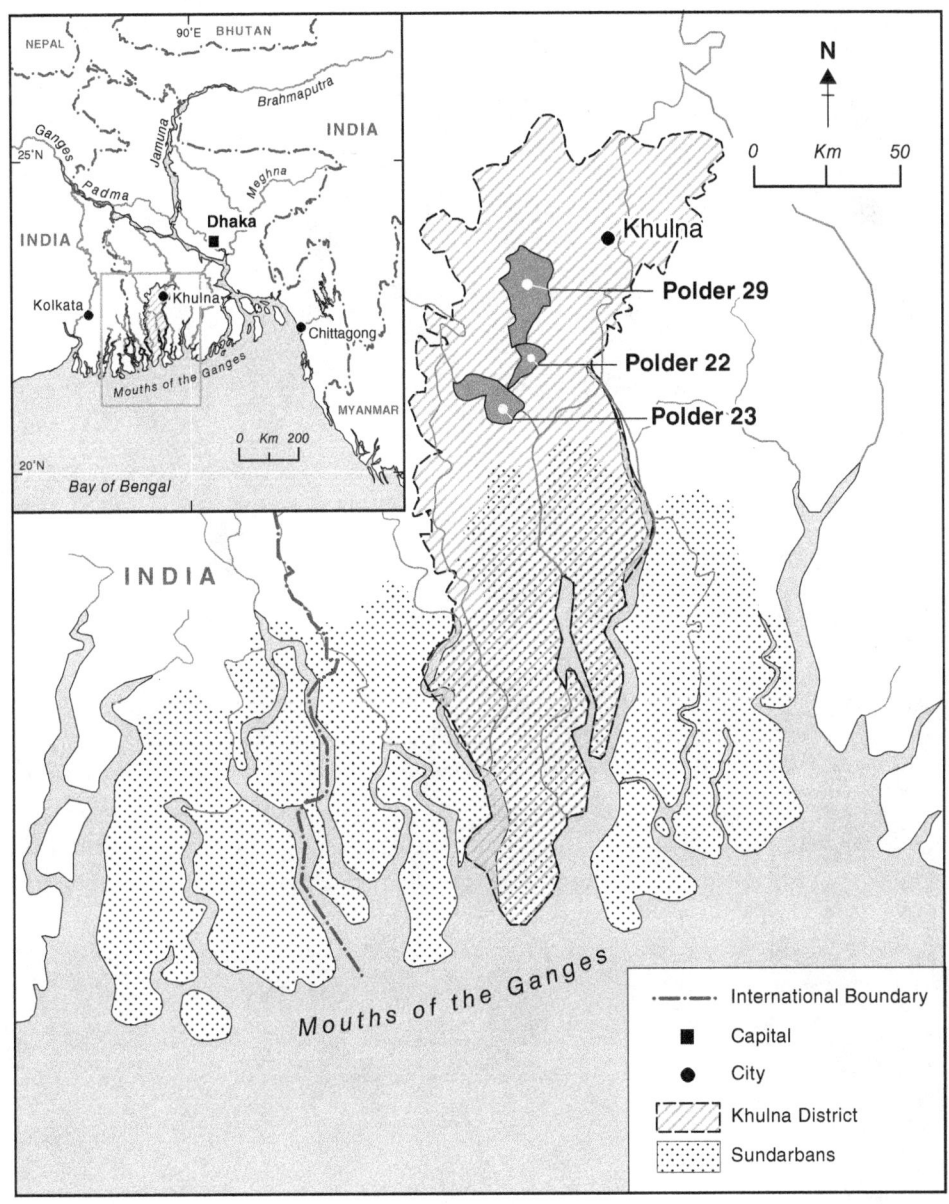

MAP 2. Map of Khulna.

Created by Mina Moshkeri.

INTRODUCTION

"They keep telling us, 'plant trees, save the environment,' but they are not stopping saltwater shrimp farming. So how will the trees survive and how will we or the environment be saved? The environment is getting polluted because of the salt water." Alpana lives in Kolanihat, a small village in the Khulna district of southwestern Bangladesh. The coastal geography of this region has been subject to extensive scrutiny, speculation, and intervention concerning its vulnerability to climate change and related ecological threats. Yet climate change is not the only threat that the region faces, and villagers like Alpana have directed their attention primarily to the social and environmental threats posed by commercial shrimp aquaculture and the development agencies that have supported its expansion.

Alpana moved to Kolanihat when she married a man from the area. At that time, her husband was an agricultural day laborer who earned a living cultivating rice in other people's fields. Her in-laws had no land of their own, but they supplemented her husband's earnings with subsistence cultivation in gardens surrounding their house, including growing vegetables and a variety of fruit trees, fishing in nearby canals, and grazing cattle on communal land and embankments. Since 1986, Kolanihat's rice paddies have been steadily taken over for shrimp cultivation, which quickly became one of Bangladesh's largest foreign exports. This transition took place through often illegal and frequently violent land grabbing, which intensified agrarian dispossession throughout the region. As shrimp cultivation spread, the lands surrounding the family's home became waterlogged year round and the soil became increasingly salinated. As a result, the fruit trees have died and the soil has become inhospitable to home gardens. The canals have all been dammed up for shrimp cultivation, and without common lands for grazing, the family can no longer raise cattle. These ecological impacts in the village have also meant that they are unable to collect drinking water or fuel for fires, thus forcing them to purchase all the items to meet their basic needs in a nearby market. The transition from rice farming to shrimp cultivation has made jobs in the village scarce, and people like Alpana's husband, who used to be employed on other people's land, are often forced to leave the village to find work.

Amid these dramatic social and ecological transformations, the livelihoods of Alpana's family and the rest of the community have become entangled in global conceptions of climate change and its governance. These conditions in Kolanihat throw into stark relief how ideas about the future actively shape the politics of the present. Who gets to imagine what the future will look like? What are the material implications of those imaginaries and how do certain imaginaries become reality while others are silenced and foreclosed? How is the power to determine the future exercised? Even as the social and ecological threats posed by climate change may be unprecedented, their impacts are far from inevitable and instead are being actively shaped in the present. Adaptation to climate change is a normative process of imagining what the future will look like and then working toward that vision, given particular conditions and constraints. The power to determine what a desirable outcome of climate change adaptation will be is the stuff of climate justice. These imaginations of the future are shaped by existing systems of power and resource distribution, which are already profoundly unequal. All strategies for adapting to climate change will benefit some people more than others. There will be winners and losers, in the words of one adaptation expert cited in chapter 6 of this book. The task of understanding climate change (what it is, what it does, and how to adapt to it) is shaped (sometimes unwittingly) by our normative commitments within these divisions. That is, existing visions of ideal social and economic structures shape our ideas about who should benefit from climate change adaptation and how. Thus, understanding these existing power structures and how they shape ideas about the impacts of climate change and strategies for adaptation is necessary in pursuing just futures in the time of climate change.

Imagining a dystopic vision of a climate-changed future for Kolanihat and surrounding areas, in recent years development agencies have begun promoting shrimp aquaculture as a strategy for adapting to climate change. Yet Alpana and other local residents have mobilized around a different vision of the future. In response to the social and ecological changes wrought by these transitions, social movements against shrimp aquaculture have sprung up throughout Bangladesh's southwestern coastal region. They demand an end to commercial shrimp aquaculture and embed this concern in a variety of other demands for agrarian justice. Many of the collectives mobilizing to stop shrimp aquaculture are organized by a national movement of landless peasants called Nijera Kori, which means "we do it ourselves" in Bengali; this movement has actively opposed commercial shrimp aquaculture for over thirty years. I have worked with Nijera Kori since 2007 and became interested in studying the agrarian political economy of shrimp aquaculture through my ongoing relationship with the group. Over the course of several years of preliminary research in the region, it became clear that today this expansion of shrimp aquaculture has become intimately tied up with the discourse of climate change and the adaptation programs that address it. Many of the ecological concerns that Alpana and others attribute to shrimp aquaculture, including waterlogging, rising soil salinity, and high rates of out-migration, have come to be referred to by development practitioners as the impacts of climate change. Development practitioners in Bangladesh increasingly demand an expansion of shrimp aquaculture as a strategy for adapting to climate change, arguing that because of the ecological threats the region is facing, it is the only viable production strategy for much of this coastal zone. These conflicting understandings of the viability of agricultural futures for the region are based on different attributions of ecological change, along with different understandings of the inevitability of those transformations. Perhaps most importantly, they are based on different understandings of desirable futures with unequally distributed benefits.

Even as these collectives in Kolanihat and beyond mobilize to put an end to shrimp aquaculture, they do not espouse a clean return to some idyllic agrarian past. Alongside their agitation for a return to rice farming, they also organize within and beyond their communities to address a variety of other political concerns. They mobilize for more equitable land distribution, for access to education and other social services from the state, and for gender inequality in labor relations and the household, including the addressing of concerns about domestic violence. All these factors together make up their vision of agrarian justice. Can they also be part of a vision of climate justice?

My focus on the politics of climate change as presented in this book is refracted through the experience of working with a social movement that does not expressly address itself to climate change. These experiences and relationships have had significant implications for how I understand climate justice. The absence of a discourse of climate justice in the movement's own narratives orients me toward a critical and generative analysis of what climate justice does or could mean in this context. I read the actually existing discourse on climate justice against the grain by unpacking the common sense of claims about the politics and possibilities for adaptation in this region. I do this by embedding an analysis of these claims within a broader global and historical political ecology of social and physical transformation in the region. I also suggest the possibility of a vision for climate justice in political movements that do not currently claim it. To that end, this work is in dialogue with other recent ethnographies concerned with climate justice that focus on communities and social movements that do not expressly address climate change (Cohen 2017; Elliott 2017; Ford and Norgaard 2020; Koslov 2016). These studies situate their understandings of climate change in relation to broader contemporary and historical struggles for social justice. Collectively, these approaches suggest new ways to expand our understanding of what climate justice can look like.

My work with Nijera Kori has given me a particular perspective both on the social and ecological changes taking place in Khulna and on the politics and practices of planning for climate change in Dhaka and beyond. In this book, I consider the ongoing production of ideas about the future under climate change generated by development practitioners and policymakers as well as the alternative visions of the future pursued by these social movements. Investigating these different imaginaries of the future, as well as the ways in which they converge and conflict, leads us to think about climate change differently: what it is, what it means for the future, how we can plan to live with it, and who will benefit. Alpana describes her own vision: "In the future it is our hope that the shrimp cultivation stops very soon. I hope people abroad stop eating tiger prawns; then we will be saved. But all the people abroad are continuing eating tiger prawns and the shrimp cultivation here continues. The rich people are after money; they don't need to care if the poor people are dying." We can learn from Alpana, especially her attention to global political economy and the associated agrarian change. She tells us that environmental justice can only be pursued through developing more equitable local production systems and making transformations in the global food regime (see also Borras and Franco 2018; McMichael 2013). While she is not explicitly concerned with climate change, her hopes for the future are deeply entangled with responses to it. The ideas of state and development agencies about the kinds of futures that are possible and desirable for her and other members of her community intimately shape what these futures will look like. Understanding these entanglements points us toward new understandings of climate justice and how to pursue equitable visions of the future.

These dynamics in Khulna must be understood in relation to broader global processes emerging in the context of climate change. By examining global discourses surrounding climate change and the policy interventions emerging in response, we find that the idea of inevitable climate crisis does not impact every community equally. In some cases, this sense of inevitability may in fact foreclose possibilities for local visions of socially just transformation. In Bangladesh, discourses of climate crisis not only obscure longer histories of dispossession, they also justify a political economy of ecological devastation through commercial shrimp aquaculture. This is the result not only of emerging global climate discourse, but also of its combination with a pervasive ideology of capitalist development that already justifies agrarian dispossession and ecological damage in the name of economic growth. The political economy of commercial shrimp aquaculture is thus central rather than incidental to this story.

Through discourses of climate crisis, Bangladesh has come to be seen as uniquely vulnerable to climate change, and therefore as a key site at which to observe its early social and material impacts. The country has thus been subject to research by a variety of actors—both academic scientists and others working for governmental and nongovernmental agencies (NGOs)—seeking to both describe what is happening and to shape strategies for response. Yet their very understandings of how these changes are taking place are shaped by these existing dystopic imaginaries and normative visions of improved futures. In turn, the landscapes themselves and patterns of environmental change that researchers seek to understand have already been fundamentally transformed through these existing systems of power. The resulting knowledge about social and environmental change in Bangladesh is characterized by a high degree of uncertainty both about what is happening and about what is driving the changes being observed. Yet the categories of certainty and uncertainty are highly unstable. Uncertainty is often *practiced* at this nexus in ways that obscure the politics of knowledge production and the normative dimensions of framing the future that are entailed in the dynamics of research and development.

While this book is focused primarily on Bangladesh, it is also a study of the political ecology of climate change adaptation much more broadly. The research for this book was undertaken between 2012 and 2017, with a focused period of two years of ethnographic research carried out primarily in Bangladesh in 2014–2015. I used a multisited ethnographic approach across multiple nodes and geographic scales in order to connect how they fit together and also to understand how power operates within and between them. This research led me from Khulna to Dhaka and Kolkata and beyond to international climate change conferences in Europe and other parts of Asia, as well as to archives in Asia, Europe, and North America. More detail on my research methods can be found in the Methodological Appendix at the end of this book.

The politics of prefiguration have perhaps never been as salient as in the time of climate change. Many facets of climate change response build on well-worn modernist teleologies of development and growth, yet, in its profound and expansive implications, climate change has also brought about unparalleled declarations of the end of history (Castree 2015; Segal 2017). Climate change, it is said, "changes everything" (Klein 2014). The sense that climate change creates imperatives that are unprecedented in human history leads to claims to similarly inexorable responses and inevitable futures.

To understand the impacts of climate change as they are experienced in particular places, we must begin with several questions about the broader political economy. How do existing production relations shape the local ecology? Who has the power to determine current practices of production and landscape management? What is the role of state and development actors in the distribution of resources? How will those resources be distributed within and between communities? The impacts of climate change and possibilities for adaptation must be studied in historically specific contexts, considering the histories of development and ecological degradation in the region, production relations among local actors, and how they shape the environment. To understand the future of the environment and possibilities for production and habitation within it, we must understand how they have been shaped in the past. The power to determine these futures is shaped by historical power dynamics and inequalities that far predate what we today recognize as climate change. Experiences of climate change are in turn mediated by these historical dynamics (Pulido 2018).

What is at issue is not whether climate change will impact lives and livelihoods. It will. Rather, at issue is how those impacts will be distributed within and between communities. These systems of distribution are not natural or inevitable. Rather, they are forged in the present and shaped by existing ideologies and power structures. In E. P. Thompson's 1975 history of enclosure and the emergence of capitalist social relations in England, *Whigs and Hunters*, he investigates how official narratives about necessary and inevitable response to contemporary ecological crisis were structured by existing and emerging capitalist class relations. While the ecology was certainly changing, so were the social relations that governed this ecology. He writes, "If we agree that 'something' needed to be done this does not entail the conclusion that *anything* might be done" (2013 [1975], 151). Precisely the same can be said of climate change. We agree that something must be done about the climate crisis. But particular plans for climate change adaptation, and for who will benefit from them, are not inevitable. They are equally shaped by existing and emerging political economies. As adaptation becomes the central legitimizing ideology of the contemporary development regime in Bangladesh, it is shaped by contestations in agrarian power relations that transcend this historical moment. These contestations shape the formulation of particular adaptation policies and outcomes.

Adaptation Regime

These power relations, how they are produced and manifested, and the contestations surrounding them are the focus of this book. I examine how they both operate through and produce a mode of governing that I call an *adaptation regime*. The adaptation regime is a socially and historically specific configuration of power that governs the landscape of possible intervention in the face of climate change. The adaptation regime evolves through the agency and interaction of multiply situated actors who collectively shape and enforce its mode of governing. Institutions of development, research, media, and science, as well as various state actors working both nationally and internationally, all participate in the adaptation regime. These actors both possess and endow the regime with authority. They legitimate this authority through their appeals to scientific knowledge about ongoing changes in the region as well as their uncertainty concerning the future implications of climate change (Watts 2015). As such, the authority of the adaptation regime is paradoxically grounded in both knowledge and uncertainty about the present and future.

The adaptation regime operates through three interrelated processes: imagination, experimentation, and dispossession. Each of these processes is produced and manifested both materially and discursively. *Imagination* refers to the work of enframing Bangladesh as a space of already existing as well as future climate crises, such that its social and ecological conditions can only be understood in relation to the impacts of climate change; the vision of future habitation of the region is similarly delimited by this sense of impending crisis.[1] This work of imagination is amplified through a process of *experimentation* with development interventions that are considered suitable for producing livelihoods appropriate to this changing climate. These interventions, which are referred to as climate change adaptation, produce agrarian *dispossession* by shaping and disciplining the possible production strategies of the region's inhabitants. This dispossession is lauded as an opportunity for development and growth, owing to its contributions to the production of export commodities. It is bolstered by the sense of inevitability of climate crisis.

To be clear, any kind of adaptation is going to constrain particular activities. This process, then, will always constitute a form of dispossession because certain practices that might produce wealth or subsistence will no longer be possible. Precisely who are the winners and losers of this dispossession defines the fundamental political economy of the adaptation regime. Each of these dynamics of imagination, experimentation, and dispossession is produced through and in conversation with existing development regimes in Bangladesh, many of which already generate dispossession or differentiation. Critically, these dynamics also characterize the development regimes that have shaped this region historically.

The adaptation regime itself does not have agency; rather, it is an agglomeration of actors (including donors, development practitioners, policymakers, researchers, and journalists) who do exercise agency within their own spheres (sometimes in parallel but often in active coordination with one another).[2] These discrete actions do not necessarily produce a coherent trajectory, yet in aggregate, they do have real, intelligible—and sometimes contradictory—effects. Therefore, in describing this regime, I do not intend to invoke a unitary entity, but rather an interconnected set of relationships that take form in a particular way in this unique historical moment in Bangladesh. As such, my descriptions of the acts of the adaptation regime refer not to the regime itself as an actor, but rather to the aggregate effects of the actors composing it. By the same token, the word *regime* does not denote a single or totalizing authority.

My understanding of the adaptation regime is both specific to Bangladesh and general to an emergent mode of global governance in the face of climate change. The adaptation regime certainly operates outside Bangladesh, and it will just as certainly manifest differently in other places. Equally, from the standpoint of different actors, the outcomes will be different. In particular, *dispossession* is the outcome of a particular context and is experienced by some actors but not all. Dispossession is fundamentally relational, both within and between communities. It is the outcome of an uneven field of social and socioenvironmental relations into which climate change intervenes. This uneven field of power underpins every element of the adaptation regime. It shapes how the climate crisis is imagined (how particular people in particular places are constituted as "in crisis" while others are not), how experiments are devised to respond to it (who decides what experiments are desirable and how they will function), and indeed what the outcome will be (who will be dispossessed and who will not).

Though my focus in this book is specific to Bangladesh, reflecting on how the adaptation regime manifests in other sites helps us to see the importance of this context and the outcomes it produces. In New York City, for example, the adaptation regime will look quite different but will share similar dynamics. Focusing on the New York borough of Queens, Rebecca Elliott demonstrates how the transformation of insurance markets in the face of climate change results in dispossession for some homeowners, governed by moral economies of "deservingness" that are deeply embedded in the social contract of the American welfare state (2017). In this context, flood insurance becomes a technology of risk governance that shapes possibilities for adaptation in anticipation of the shifting threat of sea level rise, with different outcomes for different actors (2019).[3] Liz Koslov examines how, in the nearby New York borough of Staten Island, some homeowners successfully mobilized to pressure the government to pay for home buyouts that allow them to retreat and relocate from an increasingly vulnerable coastline (2016). For these people, the state intervened to impede the dispossession that would have resulted from the existing political economy of land examined by Elliott. Yet Koslov demonstrates that not everyone in Staten Island can retreat without dispossession and that this adaptation is deeply textured by class and race (Koslov 2019).[4] In both cases, the particular context of social and socioenvironmental relations shapes the outcomes of adaptation to climate change. We see in these examples that while imagination and experimentation under the adaptation regime will transpire differently in different historical and geographic contexts, dispossession too is a political outcome that will manifest differently.

The adaptation regime concept builds on political economies of development that theorize that development regimes are intrinsically global in their material and ideological power, yet they are both administered and officially recognized in their national manifestations and particularities, with stratifying effects locally (Akhter 2015; Friedmann and McMichael 1989; Goldman 2005; Ludden 2005). These literatures have invoked the term *regime* to describe interconnected (although not necessarily unitary) modes of governing across spatial scales. While Bangladesh's adaptation regime is historically produced and concrete (as I examine later), it has been produced relationally within a global hierarchy of development and accumulation within and between nation-states (Hart 2001). I examine both global processes and local specificities to illuminate the multiple scales through which production and social reproduction are managed and governed in the name of adaptation (Goh 2019; Vaughn 2017b; Watts 2015). This analytical lens on the adaptation regime in Bangladesh is at once deeply localized and profoundly transnational, as it both shapes and is shaped by a global geopolitics of capitalist development. In this sense, it is grounded in a methodological tradition that recognizes the need to construct an understanding of global phenomena through attention to historically and geographically specific social processes (Hart 2002a, 2018; McMichael 1990). Yet even in this specificity, the adaptation regime in Bangladesh is paradigmatic, involving a variety of global actors, sites, and scales of production. Its manifestation in Bangladesh exemplifies how these global imaginaries of climate crisis shape the governance of landscapes and communities in ways that are spatially uneven and profoundly inequitable.

Bangladesh's contemporary adaptation regime is at once a continuation of and a rupture with past development regimes (Ludden 2005). In many ways adaptation programs have resulted in the same material impacts, particularly agrarian dispossession, as previous development regimes and have reinforced their longstanding logics and processes (Ireland and McKinnon 2013). Moreover, the development policy response to climate change represented by the adaptation regime draws on a long history in Bangladesh of sidelining attention to systemic issues involving power and equity in favor of technical responses to simplified crisis narratives (Lewis 2010). Yet the adaptation regime also represents a shift to a new regime of dispossession (Levien 2018) in the sense that it creates new opportunities for dispossession, legitimizes this dispossession differently and more urgently, and thus (in its claims to inevitability) poses new challenges to resistance. I explore this process of dispossession both in its imbrication with prior development models and in its novel formations. The interface of development and climate change has also produced new understandings and discourses about the landscape, how it is changing, and what must be done to respond. Many development practitioners suggest that climate change presents a break with previous strategies and logics of development, in the sense that it produces new imperatives for transformation. If the impacts of climate change are inevitable, then the mandate to adapt is also inexorable.

The articulation of prior ideologies of development with new dystopic imaginaries of climate change results in an adaptation regime that neither is entirely distinct from prior development ideologies nor incorporates their logics wholesale. In my conversations with development practitioners about the relationship between adaptation and previous development logics, I frequently encountered considerable ambiguity about the newness of climate change adaptation. Many practitioners suggested that the strategies would be the same—the promotion of shrimp aquaculture in Khulna being a prime example. But others said that climate change demanded new strategies and offered new imperatives, which they often articulated as intensifications of older strategies. For example, one practitioner talked about the need to abandon the old "rights-based approach" that says people have a right to stay in their homes if those homes are not "worth investing in." These contradictions suggest ideological formations that are neither wholly the same nor wholly different from prior development ideologies. Rather, the adaptation regime is informed by and articulated through prior development regimes.

In examining this relationship between the adaptation regime and prior development regimes, I draw on Stuart Hall's notion of "articulation," which he uses to explain how different power structures relate to one another and how new ideologies emerge out of existing ones (Hall 1980). Engaging with Gramsci's concept of "common sense," Hall explains that our systems of representation are composed of images, myths, concepts, and ideas that frame the way we understand and order the world (Gramsci 1971). "Common sense thinking," Hall writes, "contains what Gramsci called the traces of ideology without an inventory" (1980, 112); it is "the regime of the 'taken for granted'" (1985, 105). The adaptation regime comes to have power in the world precisely through these systems of representation, which already exist and are structured by the unequal distribution of power and resources within and between communities. The climate crisis that the adaptation regime predicts and enacts in coastal Bangladesh is thus already taken for granted. Deconstructing the language and behavior of this adaptation regime and the ways in which it informs depictions of climate change and its impacts on communities can help us to understand in what ways it is ideological and the systems of power it reproduces.

In theorizing about adaptation regimes in this book, I build on critical scholarship from political ecology and environmental studies concerning the new opportunities and risks associated with discourses and interventions in response to climate change (Barnes and Dove 2015; Marino and Ribot 2012; Nightingale 2017; Taylor 2015; Watts 2011, 2015). Science and technology studies provide the tools for understanding the construction of knowledge associated with such practices, producing new sociotechnical imaginaries of life in a climate-changed world (Demeritt 2001; Jasanoff and Kim 2015; Miller and Edwards 2001). These new interventions, which are conducted in the name of adaptation, are embedded in long histories of development understood as a conceptual apparatus for ordering global hierarchies of wealth and power (Ferguson 1994; Hart 2001; Li 2007; McMichael 2009). Finally, the tools of agrarian studies, which questions teleological predictions of the disappearance of the peasantry (Akram-Lodhi and Kay 2009; McMichael 2008; Wolford 2010), facilitate an exploration of what we might call the agrarian question of climate change, and how it produces difference among agrarian peoples.[5] That is, I examine the kinds of agrarian transitions that will result from climate change or from attempts to adapt to it—questions that have been debated since the publication of Kautsky's foundational text (1988 [1899]).

Dystopic Imaginaries

Threatening Dystopias offers a close examination of the process of diagnosing dystopia. Who is the subject of this power of diagnosis and who is its object? How is that power secured and exercised? What is the relationship between a diagnosis of future dystopia and the present? How does the former shape the latter? Throughout the book, I refer to "dystopic imaginaries" of climate change in Bangladesh, examining how they are formed and the kind of political work they do in the present. A dystopia is a space where everything is bad or unpleasant; the term evokes ruination, degradation, and deprivation. It also bears some relation to utopia; even as the two may be seen as obverse to one another, attempts to secure utopian visions may have dystopic consequences, and one person's utopia may be another person's dystopia (Claeys 2017). The outcomes of these epistemological tensions are not accidental; the power to diagnose dystopia is framed by existing power structures that in turn shape the nature of the utopias that will be pursued, where they will be pursued, and to whose advantage. Similar concerns are echoed in other scholarly works on apocalyptic climate imaginaries (Ginn 2015; Katz 1995; Skrimshire 2010; Swyngedouw 2010).[6] Yet unlike apocalypse, which represents a sudden rupture or break with the present, a dystopia grows out of existing social, political, and environmental realities. Apocalypse is necessarily fictional—it does not exist; this is not necessarily true of dystopia (Levitas 2010).

Imagination is also a powerful political force. It is a social practice and a means of world making. Imaginaries are not mere figments. They are both the product and the producer of real, material effects. Imaginaries "build on the world as it is, but they also project futures as they ought to be" (Jasanoff 2015, 323). In this sense, they are deeply normative—the power of imagination is linked with the politics of prefiguration. If material and intellectual power dialectically reinforce one another (Marx and Engels 1998 [1846]), then the power of imagination fortifies the power to produce the future as it is imagined. A dystopic imaginary in the time of climate change thus holds a great deal of political potency (Swyngedouw 2013a). It reflects the actually existing material realities of social and environmental dispossession, while it also prophesizes a catastrophic future collapse. This temporal ambiguity reflects a broader sense that elements of this dystopia may already exist. In its normative dimensions, a dystopic imaginary also projects an ideal alternative future. As it becomes increasingly embedded in actually existing social and material worlds, it actively shapes them. Today, this power of prefiguration in Bangladesh has coalesced into the adaptation regime, shaping ideas about the kinds of presents and futures that are seen as viable in a time of climate change. In many ways, it exacerbates the threats to which it claims to respond. The dystopic visions of a climate-changed future produced by this regime have radical implications for people who are seen as under threat. For people in rural Bangladesh, that means dispossession from communities and livelihoods that are already vulnerable.

These socioecological futures are differentiated by both class and location within Bangladesh (and beyond). What my focus on agrarian change under the adaptation regime highlights is that not all futures are rendered obsolete through these imaginaries. In the case of shrimp aquaculture, this differentiation is manifested through the dispossession specifically of landless and land-poor populations, whose livelihoods come to be seen as unviable. Elite landholders and investors in shrimp enterprises from faraway cities are seen to have a more promising future in this new adapted landscape through the expansion of aquaculture.

Similarly, the adaptation regime also differentiates between urban and rural futures under climate change. Rural communities are often thought to be more precarious and less "worth saving," in the words of one development practitioner (quoted in chapter 2), than urban communities. Yet these normative claims are also always informed by economic and developmentalist imperatives. As I examine further in chapters 2 and 3, visions for out-migration from this coastal region are shaped as much by the economic opportunities of urban growth as they are by the physical impossibility of inhabiting this coastal region. These dynamics are

FIGURE I.1. Farmer holding shrimp in Khulna.

Photo by the author.

illustrative of the emerging global politics of "retreat" in the face of climate change. Coastal retreat and resettlement are increasingly playing an important role in discussions about climate change adaptation (de Sherbinin et al. 2011). However, who should migrate, from where, the conditions under which they should do so, and with what kind of support are all matters of politics. As Koslov demonstrates in New York City, similar imperatives to those driving visions of migration from coastal Bangladesh also shape ideas about the need to continue inhabiting precarious urban coastlines in the face of climate change (Koslov 2016). In response to the demands of social movements organized to retreat from the coast of Staten Island after Hurricane Sandy, Koslov cites then–New York mayor Michael Bloomberg, who argued, "As New Yorkers, we cannot and will not abandon our waterfront. It's one of our greatest assets. We must protect, not retreat from it" (Koslov 2016, 360–361). Similarly, Marino and Shearer have written about the deeply racialized visions for retreat from coastal areas of Alaska, which exacerbate the vulnerabilities of indigenous communities impacted by climate-related erosion and sea level rise (Marino 2018; Shearer 2012). In Bangladesh, these politics of retreat are differentiated by class even as they produce difference among and between rural and urban communities.

Climate Change and Environmental Politics in Bangladesh

Climate change in Bangladesh is everywhere and nowhere. This paradox plays out in newspaper headlines, foreign embassies, the offices of the prime minister and the Planning Commission, NGOs, the World Bank, and UN agencies; at environmental protests; and in the villages on the coast of the Bay of Bengal, whose vulnerability is a source of constant speculation and intervention by those concerned with the country's development. Bangladesh is frequently referred to as the world's most vulnerable country in regard to climate change (ADB 2012; Arastoo Khan 2013; Nasiruddin and Sieghart 2014), and the World Bank calls it "the emerging 'hot spot' where climate threats and action meet."[7] Climate change has become the terrain on which Bangladesh engages with the world. It is increasingly the lens through which the nation represents itself abroad; and, in turn, it is the primary means through which the world recognizes Bangladesh. This terrain of engagement was endorsed in 2015 when the United Nations awarded Prime Minister Sheikh Hasina the United Nations Champions of the Earth award for Bangladesh's initiatives to address climate change. Conversations in the country about climate change are ubiquitous. They wind their way into topics as diverse as rice agriculture (Pinaki Roy 2014), garment manufacturing (Black 2013), microcredit (*Daily Star* 2014b), and child marriage (Human Rights Watch 2015; TakePart 2017).[8] Addressing climate change is said to be necessary for the country's economic growth (Nakao and de Boer 2015) and a means to make Bangladesh more democratic (Steele 2017) and more "cosmopolitan" (*Daily Star* 2014a).

Bangladeshis also play a major role in international climate diplomacy, having organized and led the Least Developed Countries negotiating bloc in the United Nations Framework Convention on Climate Change (UNFCCC) negotiations since the bloc's inception. Claims about climate justice and climate action are ubiquitous throughout the country's massive NGO sector. These narratives focus on the responsibility for developed countries to pay for climate action in less developed countries as reparation for their historical greenhouse gas emissions.

Yet when it comes to local political imaginaries, climate change rarely plays a role. Relative to this constant production of climate-related ideas, discourses, and interventions, climate change is surprisingly absent in local politics. It is rarely if ever invoked in local electoral campaigns, and politicians tend not to speak about it except in relation to the climate finance obligations of the developed world to Bangladesh. Activists concerned with civil and human rights rarely engage with questions of climate change and climate justice, and neither are they a significant political concern for the peasant social movements in coastal Khulna, a region that is the object of many climate change adaptation interventions (as well as the geographic center of this book). While the environmental movement within the country is quite robust, local activists remain largely unconcerned with climate change, instead devoting their energies to specifically local ecological concerns, such as open-pit coal mining, pollution from power plants and garment factories, and the impacts of shrimp aquaculture.

Nijera Kori is among these local activist groups that are concerned with local environmental issues particularly as they are embedded in broader struggles for civil and human rights. Nijera Kori is technically a nongovernmental organization (what Lewis [2017] refers to as a "radical NGO") that provides support through training and community organizing to autonomous collectives of landless people throughout the country. These collectives are composed of approximately 250,000 women and men who depend on their own physical labor as their main source of livelihood (primarily as agricultural wage laborers, sharecroppers, and subsistence farmers). These members refer to the movement overall as "Nijera Kori" or "*bhumiheen andolon*" (landless movement).[9] Thus, I use "Nijera Kori" as shorthand to refer to this movement of diverse autonomous collectives along with the organization that supports their mobilization. The organization maintains a modest central office in Dhaka, with divisional and local branches spread out around rural Bangladesh. Nijera Kori has a presence in 25 percent of Bangladesh's sixty-four districts. The work of the central office and the community organizers working in its rural branches is supported by a small group of progressive donors based in Bangladesh, Canada, and Europe.

The local civil society leaders who lead local planning and discourse surrounding climate change and are active in the construction and operation of the adaptation regime are a distinct group from Nijera Kori and other environmental activists. They work for a variety of development NGOs and university research centers. They are sometimes facetiously referred to as "the climate mafia" by donors and people in Bangladesh's development community; on occasion I also heard leaders of this group jokingly refer to themselves this way. In addition to planning, developing, studying, and advocating for climate change adaptation and finance in Bangladesh, many of these people have also been leaders in the international climate negotiations mentioned previously, in particular championing the cause of increasing climate finance to members of the Least Developed Countries bloc. Nonetheless, they remain resolutely uninvolved in these local environmental politics.

The massive protests across Bangladesh against the proposed Rampal Power Plant illuminate the contradictions between these different groups. A partnership formed between the Bangladeshi and Indian governments proposed to build a coal-fired power plant in Khulna, some nine to fourteen kilometers north of the Sundarbans, the world's largest mangrove forest. Rampal was designed to be the country's largest power plant. The project is heavily subsidized by the Indian and Bangladeshi governments, and the Institute for Energy Economics and Financial Analysis calls it "a means to sell Indian coal to Bangladesh," despite the promise of Bangladesh's rapidly expanding solar power industry (Sharda and Buckley 2016, 1). Chief among the concerns cited by Bangladeshi activists are the threats to Sundarban biodiversity, exacerbated by the inevitable increase in shipping traffic. Moreover, the land acquisition process for the plant was characterized by the violent dispossession of local residents (Mahmud, Roth, and Warner 2020; Transparency International Bangladesh 2015), including, according to one report, 400 landless families and 3,500 land-holding families (South Asians for Human Rights 2015, 10). Despite the government's professed commitment to climate action, it has rejected all efforts of activists to impede plans for this coal-fired power plant from moving forward. In 2016, the leader of the movement against Rampal, Anu Muhammad, received a series of death threats motivated by his work in the campaign; these threats were traced to the cell phone of a high-ranking member of the ruling party (Sathi 2016a). This was not the first indication of threats of violence made by the government against activists opposing the Rampal plant—another party member, a former environment minister, had previously suggested that Bangladeshi patriots might break the legs of activists who went too close to the proposed site for the plant (Sathi 2016b). Activists decry the hypocrisy of a government that professes grave concern with the impacts of greenhouse gas emissions acting so decisively to promote the burning of fossil fuels in a particularly vulnerable ecological zone. They also point out that if the Sundarbans are Bangladesh's primary defense against cyclonic storm surges, which are feared could increase in severity due to climate change, then acting to threaten this vital coastal defense is perhaps a more dangerous hypocrisy. These activists, in turn, are relatively silent on the topic of climate change itself, suggesting that other local issues are more pressing.

One evening, over a meal during the UN Climate Change Conference in Bonn, Germany (COP 23), I asked one member of the loose "climate mafia" collective why they had failed to speak out against the Rampal plant or engage at all with the movement opposing it. "We aren't activists," he told me. His point was that they do not involve themselves in local environmental politics; their conception of environmental justice is explicitly global. He seemed to be arguing that their approach is mutually exclusive with existing local visions of environmental justice. Yet, as I demonstrate in this book, the antipolitics of climate change adaptation have serious impacts on the political economy of development and agrarian change in Bangladesh—it is anything but apolitical. The contradictions between the primacy and the absence of climate change in these competing Bangladeshi political imaginaries suggest the need for a deeper interrogation of the politics and narratives of climate change both in Bangladesh and beyond. Why have these multiple scales of environmental politics have failed to find a common ground within an encompassing articulation of climate politics?

In this book, I examine the conjunctural transformations of knowledge, environment, and political economy in coastal Bangladesh in the contemporary moment, both those that claim to be linked with climate change and those that do not. In doing so, I demonstrate how the ubiquity of climate crisis narratives precipitates an inherent teleology of climate dystopia. Seeing climate change everywhere (without recognizing its interconnections with other drivers of social and ecological change) fails to appreciate the conjunctural dynamics through which a climate-changed future is actively shaped, negotiated, and contested in the present. I argue that we need to understand how diverse and discrete socioecological transformations combine and interact with climate change, as well as the implications of assuming that they are all part of the same inevitable future crisis.

Climate change is a global phenomenon with effects that are increasingly felt all over the planet. But the ways in which its impacts will be manifested in particular places are not predetermined. People will not be able to choose just what the future will look like under climate change (indeed, many quite serious climatic shifts are already locked in), but they *will* shape that future through ongoing political struggles in the present (McMichael 2008). This conjunctural analysis of the experience of climate change in Bangladesh entails a methodological and historiographic choice—to understand the contradictions, contestations, and interactions among a variety of linked and synchronous processes both historically and in the current moment—including, but not limited to, climate change (Li 2014; Wolford 2016).[10]

There is a Bengali idiom, *shak diye mach dhaka*, meaning "covering up the fish with greens," which is used to describe trying to hide something that is already well known to many. The idiom might well be used to describe a common attitude toward climate change discourse in Bangladesh, particularly among the local civil society leaders mentioned previously and the middle class more generally. While climate denialism is rare and few would disagree that the countries of the Global North should take responsibility for mitigating emissions of greenhouse gasses, there remains a general, if not often publicly articulated, skepticism about the ways in which climate change is invoked. Climate change is largely considered to be the purview of "NGOs," which literally refers to registered nongovernmental organizations but more generally invokes Bangladesh's massive development sector, which is supported through international aid from the Global North. A deep frustration with the depoliticizing impulse of NGOs and the development sector (Devine 2003; Feldman 2003) has extended to the discourse on climate change that they have forged.[11] There is a sense that the idea of climate change is deployed in ways that conceal the politics of environmental change that transpire locally.

Rising soil salinity in the southwest of Bangladesh is a case in point concerning this skepticism about the drivers of change. Though it is commonly attributed to climate change in public discourse, most in Bangladesh would instead attribute it to a variety of other well-known changes in the landscape—the construction of embankments, the diversion of much of the water of the Ganga River back to India through the Farakka Barrage, and the cultivation of shrimp (as shown in figure I.1).[12] The point here is not to argue that climate change will have no impact on coastal Bangladesh, but rather to highlight the dynamics of the contestation of knowledge production about how it will be experienced and its relationship to other drivers of environmental change. This constant production of knowledge about climate change is fraught with power dynamics that silence alternative understandings of the history and politics of environmental change in the region. That is, discourses concerning climate change in Bangladesh pivot around competing demands to recognize the role of local and global political economic dynamics in shaping the region historically, now, and in the future.

Conflicting Narratives of Social and Ecological Change

Through my ethnographic research on the adaptation regime and the environmental politics within and beyond it, I engaged with a variety of different actors both in Bangladesh and abroad. I conducted interviews and participant observation with donors, development practitioners, scientists, policymakers, activists, farmers, rural landless workers, and migrants. By working with each of these diverse sets of actors, I gained insight into how differently they understood the social and environmental challenges facing the region. These disparate understandings were manifested in different perceptions of the socioecological changes taking place, what those changes mean for the future of the region, and the normative value of those changes. These different perceptions, importantly, both were shaped by and had significant implications for material positions and actions among these various actors. Most significantly to understanding the adaptation regime, these competing interpretations of ongoing change shape the way the problem is addressed through programs for climate change adaptation.

For example, when it comes to rural out-migration from Khulna, these different actors have dramatically different understandings of what is driving migration and whether migration is a positive shift. In this case, for donors, development practitioners, and Bangladeshi government agencies, the migrations taking place out of the coastal zone are thought of as "climate migration" and considered a positive opportunity for economic growth. For scientists, their perspective on migration is contingent on their research questions, which shape the way they approach the problem. Thus, researchers who set out to study climate migration may identify it in places where those who set out to study historical patterns of agrarian transition might not. Finally, farmers and migrants in and from Khulna describe these migrations as largely the result of changes in agrarian political economy. That is, local residents see the transition from rice to shrimp as a driver of economic and ecological transformation that results in rural dispossession. They understand the changes their communities are experiencing not as sudden ruptures caused by climate change, but instead as the products of ongoing historical patterns of agrarian dispossession. Each of these groups interprets changes they observe through their own positionalities, experiences, preoccupations, and normative frameworks. Migration is a pattern that is shaped by a variety of social and material dynamics, and thus is susceptible to being attributed to any of these diverse drivers. By untangling the different drivers of change that have led to the present conjuncture, we can not only better understand the complexity of how migration is shaped by social and environmental change in the current moment but also better comprehend the political significance of these different interpretations of change.

Similarly, there is broad agreement that the region is facing significant environmental transformations, including rising soil salinity, intractable waterlogging, and declining diversity of indigenous species of plant and aquatic life. However, the causes of those changes are understood quite differently. Specifically, the role of climate change carries different weights in these different interpretations of socioecological change. Donors, development practitioners, and government agencies again largely attribute these ongoing environmental changes to climate change. For scientists, their understandings of these changes are again largely shaped by the object of their research, but for the most part they (particularly international natural scientists) believe that the current environmental changes are the result of a combination of factors, among which climate change plays a relatively minor role. Community members in Khulna attribute these changes directly to changes in rural production (specifically the transition from rice to shrimp), as well as large-scale embankment engineering projects over the past several decades.

Perhaps most importantly, these different actors hold very different perspectives on what these changes mean for the future. Donors, development practitioners, and most government agencies see the environmental crisis as inevitable, which motivates the idea that agriculture in Khulna is "no longer viable" or will no longer be viable in the near future. Clearly this sense of the inevitability of crisis has significant implications for the way that adaptation policy is designed and implemented. Scientists and farmers, however, largely see these changes in relation to particular historical and contemporary modes of planning and intervention (such as embankment engineering and aquaculture development programs). By identifying the ongoing formation and contestation of these plans and modes of intervention, scientists and residents of rural communities identify the active and ongoing production of these transformations, as opposed to their inevitability. This alternative interpretation of the inevitability of change suggests that by reimagining current policy and modes of intervention, the dystopic future facing Khulna could be avoidable.

Organization of the Book

In its broadest sense, *Threatening Dystopias* is a book about the politics of climate change, describing how its impacts as well as the impacts of attempts to adapt to it are and will be distributed now and in the future. To that end, it examines the history, discourses, understandings of, and responses to ecological change in Bangladesh today. I approach this political ecology from a variety of different geographical and historical perspectives in order to better understand the context of changes that the country is currently experiencing.

The book begins in chapter 1 with an exploration of the environmental history of the region that is now Khulna, focusing on the historical foundations of the adaptation regime. The dynamics of imagination, experimentation, and dispossession have manifested in the region from the colonial period to the present, shaping historical patterns of intervention and representation. This demonstrates not only the persistence of historical patterns of development in the region, but also how this exercise of power has shaped the landscape and the vulnerability of its inhabitants to climate change today.

In chapter 2, I focus on the politics of development interventions for climate change adaptation in Bangladesh. I do this by elaborating on the *adaptation regime*, examining the political and historical dynamics through which Bangladesh has become "ground zero" for experiments in adaptation within the new global development regime under climate change. I then turn in chapters 3 and 4 to examine the production and mobilization of scientific knowledge about climate and ecological change with a focus on different actors in the development and scientific communities. In these chapters, I demonstrate that at the local level, normative ideas about ideal socioecological futures shape the very understanding of environmental change and its possibilities. I also demonstrate how uncertainty about climate and environmental change is used to redistribute power and resources within and between communities. In chapter 3, I focus on how the politics and practices of uncertainty, as embedded in the dystopic climate imaginaries of the adaptation regime, shape scientific knowledge about climate change in Khulna. Chapter 4 shifts the focus from production to the circulations of knowledge and uncertainty about ecological change in Bangladesh and how they shape development practice in Khulna. The focus in this chapter is on development planners who act on knowledge and uncertainty about ecological change through a series of different development interventions. In both chapters 3 and 4 together, I argue that uncertainty about ecological change and its drivers and impacts in the region is claimed, produced, and mobilized to pursue particular ends within the adaptation regime. In so doing, I highlight the instability of the categories of certainty and uncertainty and how knowledge is enrolled in the production of each. Both are subject to interpretation and manipulation, and both are always in flux.

Chapters 5 and 6 bring into focus the agrarian political economy of three villages in Khulna. In chapter 5, I concentrate on a village that has transitioned entirely from rice agriculture to shrimp aquaculture, with a focus on the narratives of the residents of that village and migrants from it who have experienced that transition. I juxtapose these experiences with dominant narratives of climate migration to expose how visions of "developed futures" in the time of climate change both shape and are shaped by the production of knowledge about ongoing transformation in those communities. Chapter 6 offers an alternative vision of the future from two communities that have resisted the move to shrimp aquaculture. It examines the social mobilizations that have catalyzed this resistance and the sociopolitical contexts within which they emerged. It concludes by discussing narratives of "incremental" and "transformational" adaptation that are pursued within the adaptation regime. In so doing it highlights the importance of examining these alternative visions from communities in Khulna to think differently about what "transformation" can look like and how it is pursued. Together, these two chapters outline competing visions for rural futures being imagined by rural communities in Khulna themselves. They also highlight the diverse possibilities for the region's socioecological futures.

In the conclusion, I argue that adaptation to climate change offers the opportunity for a radical redistribution of power and resources. I examine how the landless social movements described in chapter 6 direct us toward alternative political imaginaries of climate change grounded in this vision of redistribution. I demonstrate how these movements point us toward a politics of possibility that should be the foundation of a progressive politics of climate justice. In considering these distinct climate imaginaries, which are highlighted by social movements and the adaptation regime, I argue that there are no inevitable climate futures. The impacts of and responses to climate change are always mediated by existing social and political economic structures. While the impacts of climate change are structural, they are also contingent in the sense that plans and decisions are being made in the present by actors on a variety of scales that shape how climate change is experienced now and in the future. These decisions are already actively transforming ecologies. If we recognize that these choices are being made in the present, then we can also see that climate crisis is not inevitable and alternative climate futures are possible.

1

"SLUTTISH, CARELESS, ROTTING ABUNDANCE"

Prehistories of a Climate Dystopia

> "[The Sunderbunds] are, in truth, a hideous belt of the most unpromising description, such as must cause any stranger wrecked on that coast, who should not proceed beyond the reach of the tide, to pronounce it a country fit for the residence of neither man nor beast."
>
> —J. B. Gilchrist, 1825

Long before the adaptation regime secured Bangladesh's status on the map of global climate crisis, during the colonial period, a dystopic imaginary of the region had already started forming. The region, particularly the Sundarban forest contained within it, was an object of intense fascination, anxiety, and denigration in the West.[1] Rudyard Kipling called it "unwholesome" (1922, 86), while Charles Dickens dubbed it "not healthy" (1875, 379). British geographer James Rennell, who created what is considered the first relatively accurate map of the region, opined, "I have long since forgot myself so far as to imagine that this is no part of it (the world), but only a separate part of the universe" (cited in Barrow 2003, 44).[2] Yet despite this intractable discourse of the region's remote, wild, and dystopic qualities, it has throughout recent history been enlisted in a variety of ways as a flexible resource in a shifting global political economy. This has taken the shape of both ideological and material resources, the value of which have shifted over time alongside regimes of governance. At the same time, the region's unique biophysical particularities have chronically impeded attempts to subdue it. By interrogating this contingent and particular production of nature across spatial and temporal scales, this historical accounting of the Sundarban region facilitates a better understanding of the ecological politics of the present (Hart 2002a; Loftus 2013; Sivaramakrishnan 1999; Neil Smith 2008).

In this chapter, I examine the historical foundations of the adaptation regime by tracing its elements of imagination, experimentation, and dispossession over time. I examine how each element has contributed to shaping the social and ecological life of this region since the colonial period, thus laying the foundations for the adaptation regime in the present. The chapter is organized thematically around these dynamics of the adaptation regime and their manifestations in the region through a discussion of key moments and governing events, policies, and ideologies that shaped this management. Examining the colonial period to the present illuminates two important continuities:[3] first, the persistence of distinctively colonial modes of representation in enframing this region today, and second, a colonial reshaping of the agrarian political economy that "forced the local economy to adapt to the needs of the metropolis" (Van Schendel 1982, 274).[4] These epistemic and material dynamics both laid the foundation for and profoundly shape the current adaptation regime.

In tracing these dynamics throughout history, I find not only that Bangladesh's contemporary climate vulnerability shares parallels with ecological vulnerability and attempts to address it in prior historical moments, but also that these prior regimes of landscape management have profoundly shaped the region's contemporary social and physical geography. Today Bangladesh continues to grapple with both the successes and the failures of these earlier regimes. Vulnerability to climate change in Bangladesh cannot be understood independently of either. While these interventions are part of long historical patterns, they were also contingent and not inevitable: administrators in successive development regimes repeatedly made decisions that drew on existing dystopic imaginaries of the region while prioritizing the particular economic conditions of their unique historical moment. Strategies for manipulating the land-water interface shifted over time with both available technology and shifting concerns about land improvement and stabilization. From the demands of the East India Company for more navigable waterways for their trading ships to a massive high modernist embankment network that sought to reshape Bangladesh's coast in the image of the Netherlands, the legacies of imagination, experimentation, and dispossession in the region over time haunt its contemporary inhabitants and threaten their ability to grapple with the threat of climate change.

The region's unique physical geography has influenced this dystopic imaginary. The waters of several major river systems drain into the Bay of Bengal through the Bengal Delta, including those of the Ganges, Brahmaputra, and Meghna Rivers. It is the youngest, largest, and most active river delta in the world (A. Atiq Rahman, Chowdhury, and Ahmed 2003). This is manifested in the coastal floodplain area of Khulna by the presence of a dense network of constantly shifting river distributaries,[5] which are punctuated by land masses that are in an ongoing state of erosion and accretion (see chapter 3 for a further discussion of erosion; see also Allison et al. 2003; Brammer 2014a). Thus, the borders of the coastline itself and the islands of which the coastal region is composed are naturally predisposed to shift at a rate that can be observed on an annual basis.[6] This river movement (avulsion) happens when the water flow migrates from one channel to another, thus changing the size and course of the channels as the water volume shifts. As the rivers carry alluvial sediment (which is transported from the Himalayas and also gathered along the course of the rivers), this dynamic of river movement also involves the siltation of riverbeds, which occurs when sediments are deposited, causing rivers to narrow and reduce their depth (Pethick and Orford 2013; Rogers, Goodbred, and Mondal 2013). A related feature of the region's geomorphology is that the land is constantly subsiding, as a result of both natural and anthropogenic factors (S. Brown and Nicholls 2015; D'Souza 2015; Hanebuth et al. 2013).[7] This subsidence was first observed during the colonial period but has come under particular attention recently due to its significance in measuring rates of relative sea level rise in relation to climate change (Brown and Nicholls 2015; Pethick and Orford 2013). In other words, relative sea level rise can be caused either by absolute changes in sea level (which is one result of anthropogenic global warming) or by the local vertical movement of land related to local and regional geomorphological processes.

The fluidity of this landscape came to be seen as both a threat and an opportunity by the British. Environmental historians of Bengal have demonstrated how the East India Company and later the British Raj grappled with "the paradox of permanence in a mobile landscape" (Debjani Bhattacharyya 2018a, 242) through experiments in both the physical and legal engineering of land and water.[8] While large embankment infrastructures attempted to fix the shifting courses of the river channels, lawmakers also devised new legal mechanisms to stabilize fixed property regimes in a geography in which land was in a constant state of emergence and dissolution (Debjani Bhattacharyya 2018a, 2018b; Lahiri-Dutt 2014; Lahiri-Dutt and Samanta 2013). While the landscape posed challenges to predicting and calculating revenues and expanding cultivation, it also created opportunities for asserting the eminent domain of the state over new lands as they appeared. This in turn paved the way for a transformation of agrarian production and labor regimes throughout the region.

Tracing these dynamics of the contemporary dystopic climate imaginary back through the colonial period is analytically significant in two ways. The first is in illuminating a longer history of contemporary ways of seeing and intervening in the landscape. Current discourses surrounding Bangladesh's unique vulnerability to climate change are not new but rather draw on historical tropes about the region's inherent biophysical vulnerability due to its low elevation, geomorphological instability, cyclonic activity, and mangrove ecology. Moreover, tracing this history helps us to understand that the current ecological crisis faced by the region has been profoundly shaped by these modes of intervening in and imagining it. In particular, we can see that even in the face of dystopic imaginaries, attempts to transform the landscape to make it more governable and amenable to profit extraction have themselves exacerbated this vulnerability. Indeed, the vulnerability of southwestern Bangladesh to climate change is the product of the imbrication of its particular local ecology with global circuits of capital and governance (Hart 2002b; Watts 2003). Examining this history is thus essential to understanding the political ecology of the adaptation regime.

Waste

Both the state and private capital (particularly landlords and moneylenders) have played a critical role in the exploitative property relations that have shaped patterns of land conversion, investment, and the control of agrarian labor. In 1793, the East India Company introduced a new land revenue system in Bengal through the Act of Permanent Settlement. The Permanent Settlement was a major administrative intervention aimed at imposing a western notion of private property (Ranajit Guha 1982) and was a key tool in nineteenth-century state making in Bengal (Sivaramakrishnan 1999). In exchange for the British renunciation of any future increase in land taxation, the Permanent Settlement granted permanent property rights to a select group of Bengali *zamindars* (large landholders) who committed to large, fixed cash payments to the Raj. These zamindars were in turn responsible for collecting rents directly from cultivators who worked in their estates, thus creating a new landlord class with almost absolute authority over the land and agrarian economy (Iqbal 2010; Marshall 1987). Significantly, the codification of these property rights also necessitated the designation of a category of "wastelands," which were not yet cultivated and therefore could not yet be subjected to private taxation (Sivaramakrishnan 1999). The result was two linked but differentiated categorizations: between groups of people related to the land (landlords and cultivators) and between forms of the land itself (wasteland and other forms of land subjected to different taxation regimes). Wastelands were not included in the Permanent Settlement with the zamindars and instead were considered British colonial government property. The lands were leased to the zamindars, who in turn leased them to subsidiary landholders or sharecroppers (*jotedars* and *borgadars*), who performed the work of reclamation and settlement (Mukherjee 1969b). The Sundarban region, which was covered with dense mangrove forest and was largely uncultivated, was thus officially designated as "wasteland" while all of Bengal was being brought under a new regime of private property rights (Pargiter 1934).

However, the Act of 1793 was passed with no land survey or method of assessing or recording land rights (Ludden 2011). As these methods were developed over the remainder of the eighteenth and early nineteenth centuries, a new Office of the Commissioner of the Sundarbans was established in 1816 (Beveridge 1876a).[9] The commissioner's primary task was to promote land reclamation (a topic explored in further detail later in the chapter) in order to facilitate the expansion of cultivation and thus turn "waste" into land that generated tax revenue (Iqbal 2010; Sarkar 2010). Administrators believed that the instigation of private property rights would create an incentive to expand the areas under cultivation by motivating the improvement (meaning clearing and cultivation) of wastes (Ranajit Guha 1982). As this cultivation expanded, new leases were granted, the revenues from which were managed by the commissioner of the Sundarbans. The material and ideological conditions that both gave rise to and resulted from this designation of the Sundarbans in particular are explored in further detail later in the chapter.

For the British, land was technically considered waste if it did not generate tax revenue for the Raj (Ariza-Montobbio et al. 2010). The designation of wastelands facilitated enclosure (Goldstein 2012), rendering the land an object of "improvement" (Ranajit Guha 1982). Yet the category was applied to a wide variety of existing ecological and biophysical conditions and to areas with extremely heterogeneous uses (Singh 2013). Thus, the Sundarbans and other areas that were considered "jungle," another heterogeneous category (Sivaramakrishnan 1999), were designated as waste along with many common lands such as pasture within villages that were used for grazing cattle (de Hoop and Arora 2017).[10]

The discursive construction of certain lands as "waste" did political and ideological work for the empire (Gidwani 1992).[11] In addition to having a revenue imperative (leasing wastelands to be brought under cultivation generated increased revenues for the colonial state), it also facilitated the British justification for colonization by simultaneously emphasizing what they understood to be the cultural and physical inferiority of Indians to the British. The British thus encouraged the development of wastelands for cultivation, offering new opportunities for governance and accumulation (Greenough 1998; Ramachandra Guha 1990).[12] The categorization of waste thus laid the foundation for material interventions that facilitated extraction and accumulation by the colonial state and private British investors and corporations (Baka 2013, 2017; Gidwani 2012; Gidwani and Reddy 2011; Goldstein 2012; Isenberg 2016). The government granted special leases to zamindars for cultivation,[13] with subleases granted by the zamindars to cultivators (Ludden 2011; Mukherjee 1983; Pargiter 1934). The government itself retained the right to determine the land's potential for productivity. The colonial imperative of expanded capitalist accumulation along with its normative conceptions of the value of land (and the people and production relations inhabiting it) combined in wasteland development discourses as the foundation

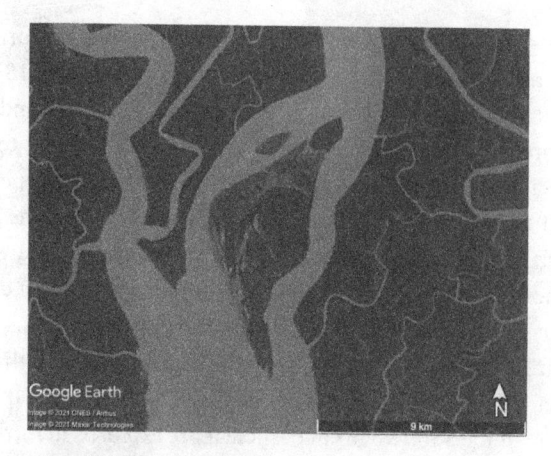

FIGURE 1.1. Example of char formation in the Sundarbans; image taken in 1984.

Source: Google Earth.

for governing both space and people in the Sundarban region. These factors continue to shape the way in which the region is governed today.[14]

In the Sundarban region, the extent and variety of the lands that were categorized as waste offered ample scope for colonial administrators to devise methods of "scientifically," politically, and physically managing the landscape. As one administrator observed, "This wilderness and labyrinth of rivers [was] the property of no landholder, but of the [colonial] sovereign," and thus cultivation and extraction could be undertaken on the basis of this sovereignty over wasteland (Hamilton 1820, 126). In addition to being home to extensive forested areas (which made up the majority of wastelands throughout British India), the *chars* and *diaras* (alluvial sediment deposits that gradually develop into new land masses) were also classified as wasteland (Lahiri-Dutt and Samanta 2013). Figures 1.1 and 1.2 depict an example of the gradual formation of such alluvial land masses in the Sundarbans in the last several decades. The classification of the Sundarbans as waste was also claimed by colonial administrators as cause for excluding the region from the Permanent Settlement, thus establishing the government's indefinite right to profit from all resources extracted from the area (Iqbal 2010; Pargiter 1934).

Accumulation

If the Sundarbans have been framed in colonial and postcolonial imaginaries as wild and remote, their position in global political economies has been anything but marginal. The region's unstable ecological and biophysical features have in

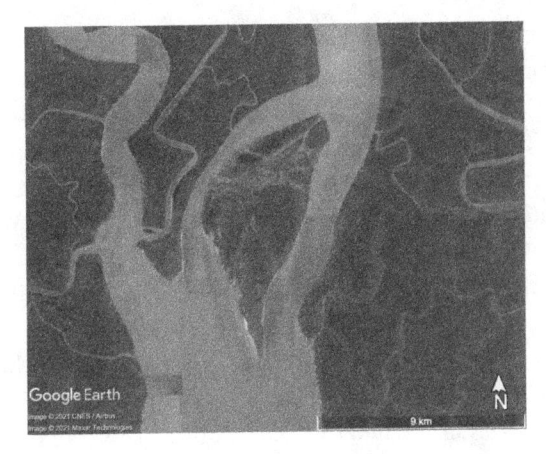

FIGURE 1.2. Example of char formation in the Sundarbans; image taken in 2020.

Source: Google Earth.

many ways undermined attempts at governance and accumulation. As successive development regimes have sought greater control over the landscape, these environmental challenges have stoked dystopic imaginaries of the region.

The British East India Company established its primary port and trading base in Kolkata (then Calcutta) in the late seventeenth century, and Kolkata subsequently became the capital of the British Raj. The Port of Calcutta became the most important port in British India, with all shipping traffic routed through the Sundarbans (Hunter 1875a, 16), leading one colonial observer to refer to the Sundarbans as "the British emporium of the East" (Bull 1823, 124). The growing significance of these shipping routes contributed to the importance of Calcutta as a commercial city and increased the power of British political and economic control in India (Mukerjee 1938, 26). The 1908 *Khulna Gazetteer* referred to these shipping routes as "one of the most important systems of inland navigation in the world," with trafficked goods valued at nearly 4 million pounds sterling annually, equivalent to about US$600 million today (O'Malley 1908, 128). The town of Khulna, located just north of the forested area of the Sundarbans, was an important node in this network, serving as the headquarters of the Salt Department under the East India Company and the "grand mart for all Sundarbans trade" (Hunter 1875b, 222) well into the twentieth century.

As Great Britain's economic power expanded in the subcontinent, so did the volume of British trading ships and the traffic through the Sundarbans. However, the constant movement of the rivers, shoals, and chars presented serious challenges to this ship traffic. In a series of six lectures delivered in Calcutta in 1906 entitled "Waterways in Bengal: Their Economic Value and the Methods Employed for Their Improvement," one government engineer complained of the large amount of money that British steamer companies were forced to write off every year due to shipwrecks in the Sundarban passages (Lees 1906, 26). This predicament led administrators on an unabating quest to discover ways to reshape and control the waterways. In fact, the initial objective of Rennell's surveys in the eighteenth century (which ultimately produced his atlas of the Delta, as described in note 2) was to map the shipping routes to Calcutta in order to facilitate the movement of goods and more efficient revenue collection, as well as the transportation of colonial troops (Barrow 2003; Rennell 1910). At a forum held by the Society of Arts in London, one British official appealed for greater efforts to develop the Sundarbans passages in a manner similar to other major world economic powers, noting, "America and Germany had utilized and improved their rivers, and made them capable of carrying an enormous quantity of goods at a very cheap freight, and the trade of those countries in consequence had prospered to an enviable degree" (Buckely 1906, 434). Thus, proposed remedies for challenges of navigation through the Sundarbans were abundant. As this same official explained, "A little deepening of the sand banks on the Ganges and the Brahmaputra would mean that all the steamers would be able to carry several hundred tons more than they would otherwise be able to do. What that would mean to the traffic of the country and to the rates charged was obvious" (434). These proposals for re-engineering the landscape have been a recurring concern of subsequent development regimes in the region (e.g. International Bank for Reconstruction and Development 1972c; United Nations Water Control Mission to Pakistan 1959) and will be taken up further later in the chapter.

As the economic priorities of the region's development regimes shifted, so did the governance of the Sundarbans. In the late nineteenth century, colonial planners began to see the development of a railway network across the subcontinent as integral to the progress and expansion of the British empire. As Iftekhar Iqbal has explored in his environmental history of the Bengal Delta, the rail network increasingly took priority over the navigability of the vast network of waterways (Iqbal 2010). In a pamphlet entitled "A Letter to the Shareholders of the East Indian Railway, and to the Commercial Capitalists of England and India," the author argued that expanding the rail network into the Delta from Calcutta would be of "universal benefit" to Britain, as "Calcutta is your emporium" and "the Ganges Valley is your manufactory—your trading ground—your source of wealth" (Transit 1848, 8). Thus, facilitating the extraction and transport of resources from the delta region was necessary to the expansion of the entire empire. As explored later in the chapter, the governance of forest resources in the Sundarbans also shifted as administrators recognized that building this vast railway network would require a substantial amount of timber and that the sundri mangroves were exceptionally well suited for the purpose, as they were used in the construction of railroad tracks (Buckland 1901; O'Malley 1908).

After the Partition of India, Khulna continued to hold a significant place in the developmentalist imaginary of expanded accumulation in Pakistan. Partition's aftermath was characterized by the political and economic colonization of East Pakistan by West Pakistan (Feldman 1999; Van Schendel 2009; Wood 1981). This took the shape of intensive resource extraction from East Pakistan by West Pakistan, along with the failure of the utopian pre-Partition promise of agrarian reform (Blair 1978; Hashmi 1992; Lewis 2011).[15] East Pakistan systematically subsidized the industrial growth of West Pakistan, largely through the export of raw jute (a natural agricultural fiber used for making rope and burlap) (Sobhan 1962, 1971). Tensions between the two halves of the country grew through the intensification of what the East perceived as economic exploitation by the West. For example, in spite of the fact that East Pakistan earned approximately 60 to 70 percent of the country's foreign exchange in the decade after Partition (Sobhan 1962, 37), it received only 20 percent of the country's development expenditure (Haq et al. 1976). While development investments were nationally concentrated in the growth of industrial manufacturing (A. M. Huq 1958), the largest investments in East Pakistan were made in water management infrastructures in Khulna (International Bank for Reconstruction and Development 1972b). These investments were justified on the basis of the goals of expanding the cropping area and growing yields in order to facilitate increased exports of raw agricultural resources (Warner 2008). The results of these interventions and infrastructural investments will be explored later in the chapter.

Although the priorities of these successive development regimes can be understood in relation to economic imperatives and potential for accumulation, they should also be understood in relation to broader concerns related to governance and state building. Since the colonial period, governance objectives in the region have integrated, on the one hand, resource exploitation and capitalist accumulation, and on the other hand, priorities related to state territoriality and the production and maintenance of hegemonic legitimacy. Barrow argues that Rennell's *Bengal Atlas* served this purpose of extending the legitimacy of British East India Company rule in Bengal (at a time when it was under threat and particularly weak), writing that Rennell "gave Company rule a colonial character, suggesting that the Company was interested in governance and improvement and not just in conquest" (Barrow 2003, 40). After Partition, aid flows were motivated strongly by the competing interests in the Cold War, both of which saw investments in development as vital in establishing ideological legitimacy in the region (Sobhan 1982; Van Schendel 2009); in particular, agencies from both the West and the

Soviet bloc invested heavily in water management infrastructure in the region that is now southwestern Bangladesh (Agency for International Development 1967).

Temporariness

The impermanence of this landscape has consistently been among the greatest challenges to those seeking to govern it, a challenge that characterized these successive development regimes. Since the colonial period, the work of transforming Bengal's rivers into an economic resource was understood by administrators to be contingent on "rationalizing" a river system that was in a constant state of flux (D'Souza 2015). The ephemeral and unstable waterways have in turn resulted in temporary political geographies. Recognizing this relationship leads to a better understanding of the dynamics that gave rise to the contemporary adaptation regime.

The natural dynamics of river movement and deltaic transformation were a source of unremitting anxiety for colonial administrators, who questioned whether the land in the Delta was stable enough to inhabit or to accommodate the kinds of durable settlements they envisaged. One colonial administrator, who was chronicling the effects of these river movements on landholders, remarked, "No buildings intended for duration can be raised on so unstable a foundation" (Hamilton 1828, 175). These questions about whether the land was fit for habitation resonate today with contemporary concerns about whether climate change will make coastal Bangladesh uninhabitable, a narrative that is deeply embedded in discourses of the adaptation regime.

The British sought at length to understand and document the constantly shifting nature of the landscape as well as to find scientific means to mitigate its transience. One former judge in the Calcutta High Court recalled,

> I can state that, between the years 1842 and 1869 inclusive, or my period of active service in India, I never recollect a time in which some proposal or other was not under discussion, in the Press and in official correspondence, in regard to these rivers. There was, generally, a dread that this or that channel was silting up; and there were repeated proposals for dredging, improving, or widening the channels. (Seton-Karr 1899, 651)

In addition to these concerns about how to physically manage the river channels, colonial officials also deliberated extensively about how to manage the rights to alluvial accretions and their associated economic benefits. While the temporariness of the land threatened the stability of colonial accumulation, colonial administrators sought to manage the temporary geography of deltaic land masses by devising new means of legally recognizing the economic value of unexpected possibilities for accumulation. This was carried out through unique land tenure arrangements that temporarily suspended taxation to promote expanded cultivation as well as physical interventions to reclaim land mass, both of which are explored in greater detail later in the chapter.

Imagination

The imaginative geographies of catastrophe and dystopia that characterize the adaptation regime far predate it. Both colonial administrators and East Pakistan–era development planners described the natural threats facing the region as "existential," meaning the very continued existence of the physical space and the communities inhabiting it was imperiled. They also considered the causes as well as potential time frames of these immanent threats to be uncertain. Moreover, the nature of these imaginaries is deeply shaped by normative conceptions of the lives of people inhabiting these landscapes.

In describing the Sundarban region in both official documents and popular texts, colonial officials described rural life with great linguistic flourishes, using words such as "dreary," "gloomy," "desolate," "miserable," "depressing," "miasmatic," and "umbrageous." These accounts of the Sundarbans often blended imperial and scientific authority with the exaggerated effect of dystopian fiction. H. James Rainey, a British subject writing from Khulna, published one such fictionalized account in the Calcutta literary journal *Mookerjee's Magazine* describing the apocalyptic end of the imaginary coastal city of Bangálah:

> Let us imagine the last day of Bangálah,—the utter annihilation of a populous city. How was this over-whelming calamity brought about? Did it sink beneath the surface of the dark waters of the Bay, amid the convulsions of nature? That page of history which ought to have recorded such an appalling event, is, at least as far we are aware, a perfect blank, so we think we may be allowed to fill it up as we best can. (Rainey 1872, 348)[16]

Rainey continues in detailed, theatrical narration to describe the "shreaks [sic] of a hundred thousand frantic souls" perishing in a violent cyclonic storm surge: "Sunk underneath the Indian flood!" The story ultimately fizzles out in muddled speculation about the fate of the "Muhammadan" inhabitants of the coast at the "gates of high Heaven" who were "ill prepared indeed to face the Eternal" (Rainey 1872, 348). Here Rainey overtly equates a deeply Islamophobic orientalism with a colonial ecological imaginary that also regarded the Sundarban climate and ecology as "evil" (D'Souza 2015). While Rainey's environmental fable was plainly fictionalized, it reveals the affective qualities of colonial dystopic imaginaries. The imaginative force of these narratives encouraged colonial attempts to control and subdue the physical and ecological landscape (and to extract from it). To the extent that these imaginaries fortified such efforts to intervene in the landscape, which in turn resulted in greater threats to the environment and its inhabitants, they supported the production of the very future to which their anxiety was oriented. These imaginaries resonate powerfully with contemporary accounts of climate crisis, in particular the narratives I describe in chapter 2 as "climate crisis memoir."

Environmental threats coalesce in these imaginaries with more general misfortunes and failings understood to characterize the lifestyle of rural coastal inhabitants. Thus, one former inspector-general of forests wrote that "to live in a boat in the Sundarbans is charming, for you have the means of escape under your feet; to live on land must be horrible in the extreme" (Eardley-Wilmot 1910, 236). As these accounts of coastal life unfold, the reader finds that subsidence, cyclones, and other biophysical risks are lumped together with imaginaries of sociopolitical hazards such as the threats of Portuguese and Arakanese pirates and of "Jungle Fever" (Mackay 1860; Mukherjee 1983; Phillimore 1945; Rainey 1868).[17] The frequency of disasters and imagination of environmental catastrophe seem to be inseparable from these concomitant dangers.

Moreover, fearful and derisive reflections on the landscape are blended with aspersions on the moral fortitude of its inhabitants and their cultivation of it. The failure to subdue the catastrophic environment of the Sundarbans is attributed to the "indolent and improvident habits" (Hamilton 1828, 183) of its "heathen natives" (Anonymous 1859, 19), their "unambitious" disposition (Mundy 1858, 283), their aversion to "even the semblance of innovation" (Hamilton 1828, 189), and their possession of "neither the means nor the intelligence necessary" to improve the landscape or its cultivation (O'Malley 1908, 100). Thus we find that the imaginative geographies of dystopia in the Sundarbans are shaped as much by the orientalism of the colonizers as they are by a study of the biophysical dynamics of the environment itself.[18]

A paradox woven throughout these imaginaries is that they fluctuate between embracing these dystopic qualities as both a threat to and an opportunity for accumulation in the region. One recurring trope involves discussions of the fertility of the soil, including normative claims about its implications and virtues. On the one hand, this fertility is acclaimed for what the colonizers seem to regard as limitless potential for extraction. One British conservator of forests wrote of the Sundarbans that "reproduction is most favorable. . . . It has been put forward that reproduction all over the Sunderbuns is unlimited, and that cleared blocks will be covered again with forest in a very short time" (Schlich 1876, 9). During certain time periods, this unlimited potential was taken as a justification for unrestricted deforestation and resource extraction from the region. However, at the same time, this fertility presented a challenge to colonial administration of the land and its resources. One official demonstrated this paradox in remarking on the villages of the Sundarbans region through vivid descriptions of their "sluttish, careless, rotting abundance" (Seton-Karr 1883, 424). Even as they praised it for its promise in greatly expanding the taxation base of the Raj, administrators lamented this fertility as "excessive" (Rainey 1891), even "evil" (Westland 1874, 178), requiring careful and constant management in order to be brought under controlled cultivation. The proclivity of reclaimed lands to relapse rapidly into untamed jungle undercut their attempts to subdue expanding Sundarban estates for the planned and managed cultivation of rice.

Experimentation

In response to these dystopic imaginaries, successive development regimes have devised a variety of technical and social experiments to transform the landscape and the communities that inhabit it. The focus of much of this experimentation from the colonial to the East Pakistan period to the present has been on building embankments that have profoundly transformed the landscape. Yet these experiments in landscape engineering have rarely been accompanied by realistic long-term strategies for maintenance. Their subsequent failures are among the primary factors shaping the region's climate vulnerability today. In this section, I focus on the dynamics of this experimentation in two particular moments: first, in the colonial period (dating to the experimental visions of the East India Company), and second, in the Coastal Embankment Project in the mid-twentieth century, which constructed a massive system of Dutch-style polders across the coast. I examine both these experiments and their implications in turn in the following discussion.

The uncertain threat of environmental crisis has consistently been evoked as the specter demanding these experiments. As one English merchant wrote,

> Every thing and every one must be prepared to see a day when, in the midst of the horrors of a hurricane, they will find a terrific mass of salt water rolling in, or rising up upon them with such rapidity that in a few minutes the whole settlement will be inundated to a depth of from five to fifteen feet! unless it be duly secured against such a calamity by efficient bunds, say of 20 feet high.... Such a visiting may not occur for the next five years, or for the next twenty years; but it may occur in the coming month of October. (Piddington 1853, 20)

These "bunds," or embankments, then, have consistently been proposed as possible safeguards against potential future destruction. Designs for such protective infrastructure have drawn on a broad range of engineering technologies and scientific hypotheses about effective management, while questions have also consistently been raised about the suitability of these technologies to local conditions. Experimentation with embankments in the Sundarbans served a dual purpose: to protect the land from inundation and also to exercise greater control over water and sediment, the movement of which caused the river channels to shift (which impeded ship traffic within them).

During the colonial period, the construction and maintenance of embankments was a constant preoccupation of administrators concerned with governing the Sundarbans and was employed for the purposes of both protection and appropriation. Beginning in 1770, the East India Company began building embankments in the Sundarbans for the purposes of artificially building up land mass through the accretion of sediment in order to expand the area under cultivation (Bandyopadhyay 1987; Maitra 1972).[19] This work of land reclamation resulted in rapid deforestation and expansion of the area of settlement and cultivation. The 1908 *Khulna Gazetteer* thus stated that "the forest is being replaced by smiling rice fields" (O'Malley 1908, 2) and that "cultivation and villages now exist where a century ago all was waste" (O'Malley 1908, 3). As these marshy lands were embanked and drained through protection from tidal inundation, the boundaries of the forested area of the Sundarbans gradually receded southward.

This reclamation took place through active promotion by the government. Land settlement and taxation laws were written specifically for this purpose, creating incentives for bringing the wastelands under cultivation and penalizing tenants for not bringing their entire plots under cultivation within fixed periods. Renting out land as Temporarily Settled Estates allowed for the government to extract greater rents from tenants over time, unlike lands within the area of the Permanent Settlement (for which revenues were fixed). For example, between 1882 and 1901, the land revenues of Khulna almost doubled, an increase attributed primarily to the enhanced rents from reclaimed land in Temporarily Settled Estates in the Sundarbans (O'Malley 1908, 149). The majority of these leases of uncultivated lands were granted to British subjects, owing to their "superior knowledge" (Gokul Chandra Das 1996, 58), as well as to high-ranking Indian mercantilists with positions in the Revenue Department of the Raj.

Needless to say, these urban elites who were granted leases in the Sundarbans did not themselves undertake the task of land reclamation. The labor required for this work of reclamation was significant as well as arduous, involving manually clearing the densely forested jungle for planting. Administrators often sought to secure the labor of *raiyats* (peasant laborers) who were already locally employed in government salt manufacture (which did not require land reclamation) (Westland 1874, 68). It thus both depended on and entrenched class differentiation. In the late eighteenth century, a scheme was even floated to expand the cultivated area of the Sundarbans through grants of wasteland to convicts (63).

One Indian geographer estimated in 1969 that in the preceding two hundred years, about half the 20,950 square kilometers of Sundarban forest had been prematurely reclaimed (Mukherjee 1969a, 311).[20] If the environmental impacts of this reclamation were not well understood before it began, they quickly began to make themselves apparent. As the reclaimed area expanded, the tidal flow in and out of the delta was cut off (Westland 1874, 180), accompanied by a cascading series of environmental threats. The river channels began to silt up, meaning that alluvial sediment, having been deprived of space to flow onto land during tidal inundation, was instead deposited on the bed of the rivers and canals, which slowly became narrower and shallower (Bandyopadhyay 1987).

As the channels narrowed, they became hydraulically "unfit," meaning the water becomes so concentrated in the remaining space that it puts excessive stress on the embankments, which periodically breach as a result (Maiti, Das, and Majee 2010, 25). Meanwhile, in the absence of the sediment deposits that formerly built up the land within the embankments, the contained area gradually subsided (Jnanabrata Bhattacharyya 1990; Tapan Kumar Das and Maiti 2010, 25).[21] With the river beds rising and the settled land inside the embankments falling, drainage became more difficult as the high tides (and even sometimes the low tides) were higher than the land, resulting in waterlogging where there was insufficient space for water (either from breached embankments or rains) to drain back into the rivers through gravity (IOR 1915, 3). All these biophysical phenomena resulting from embankment construction continue to play an important role in the physical (and also the social) landscape of Khulna.

This dynamic of using embankment technologies to artificially reclaim land in the Delta continued during the East Pakistan period. It was entrenched through the creation of the East Pakistan Water and Power Development Authority (EPWAPDA), an autonomous government agency that by the 1960s was receiving approximately 20 percent of East Pakistan's development resources, an estimated 15 to 20 percent of which was paid directly to foreign consultants (John W. Thomas 1972c, 8). EPWAPDA projects to develop new water infrastructures throughout East Pakistan embodied the donor-driven, high-cost, technically complex infrastructure design that characterized postwar high modernist development planning. These designs also diverged further from local water and land management regimes than ever before. One USAID consultant wrote in a scathing evaluation of their programs that

> Engineers working on every one of [EP]WAPDA's projects under valued the importance of communicating with farmers to gain their support, organizing them for irrigated farming, and assisting them in learning to produce under new conditions. Too often the solution to agricultural problems was to go ahead with design that suited engineering requirements and assume that a call for more agricultural extension officers or technical assistance will solve the problems. [EP]WAPDA projects adequately proved that this easy formula was totally inadequate. (John W. Thomas 1972c, 30)

In Khulna, this inadequacy was largely manifested in the biophysical phenomena (described previously) resulting from embankment engineering interventions. The failure of these programs in design and maintenance, as well as the inadequacy of the response to the recognition of these failures, comprise what may be the greatest physical challenge to coastal communities in Khulna today.

The most significant of these projects undertaken by the EPWAPDA was the Coastal Embankment Project (CEP). The existing embankments, which had been built prior to independence, had been rebuilt annually, allowing for seasonal inundations that facilitated the periodic deposit of sediments in the tidal floodplains. This maintenance was supported by zamindars, using the labor of the cultivators. After the collapse of the zamindari system following Partition, the embankments deteriorated (Leedshill–De Leuw Engineers 1968, 18). The CEP was developed not only to restore the embankments, but also to establish an entirely new water regime in the coastal region. Initial plans were developed by the International Engineering Company (IECO), a San Francisco–based firm that developed several major infrastructure programs for EPWAPDA through financing from USAID (Agency for International Development 1967), along with additional engineering consultants from another San Francisco–based engineering firm, Leedshill–De Leuw (Leedshill–De Leuw Engineers 1968). The latter wrote in their initial report that the project "will rank among the largest undertaken by Pakistan and will rank high on the list of earthmoving projects in the entire world" (155). Initiated in 1961 at an estimated cost of $55 million, the plans, time frame, and costs of the project repeatedly swelled, such that by 1971, estimated costs had grown to $278 million (US General Accounting Office 1971).[22] In planning reports, foreign consultants repeatedly invoked the "existential" necessity of this undertaking to East Pakistan, suggesting that the physical existence of the province and the survival of its population were at stake (National Research Council 1971; United Nations Water Control Mission to Pakistan 1959). At the same time, however, they also recognized the potential impact of the CEP on the US economy. One USAID report justified the loans to Pakistan for the CEP candidly,

The loan funds will be used for the purchase of services (and possibly a small amount of goods associated with these services) in the U.S. U.S. firms and experts will be employed. Local currency costs of contracts will be met by the GOP—thus there will be no direct outflow of dollars. The effects of the loan on the balance of payments initially will be neutral. . . . In the long run, the loan will have a favorable effect on the balance of payments, because the full amount of the loan including interest will be repaid to the U.S. in dollars. (Agency for International Development 1967, 38)

If the economic motivations of the project for the consultants (who designed and then were paid to manage the construction of the embankments) were clear, the impact on the coastal region and its inhabitants was even more dramatic. The CEP was designed to protect coastal lands from tidal inundation in order to expand the cropping area for increased food production. Planners estimated that on completion, the approximately 2,000 miles of embankments would provide protection for 3.4 million acres of fertile land across four coastal districts (including 1.26 million acres in Khulna alone) (Bari 1978, 18; International Bank for Reconstruction and Development 1972b). The essence of the plan was to build a system of 108 polders (today there are a total of 123) across the coast. "Polder" is a Dutch word for a low-lying tract of land completely surrounded by a protective dike; the infrastructure is intended to artificially isolate the polder from the adjacent hydrological system, allowing internal water levels to be controlled through mechanical pumps and sluice gates. New agronomic conditions within the polders would allow for the expanded cultivation of new high-yielding foreign rice varieties (International Bank for Reconstruction and Development 1972c). By creating these larger land units and eliminating the movement of water through tidal streams within them, the polder system was designed to ultimately close off the coastal estuaries (International Bank for Reconstruction and Development 1972d).

However, problems with the technical design of the polder system manifested almost immediately. The embankments cut off the flow of water that was necessary for agriculture in many parts of the polders, while causing waterlogging in other parts, where they inhibited drainage (Advisory Group on Development of Deltaic Areas 1966; Leedshill–De Leuw Engineers 1968; National Research Council 1971, 38; Saifuzzaman and Alam 2010). In some areas, faulty designs exacerbated saline intrusion and siltation, as opposed to preventing them (Bari 1978).

These problems highlighted in particular the failures of local consultation and lack of understanding of the complex delta environment among the foreign consultants who were employed in project design (John W. Thomas 1972a). One researcher likened the attempt to transfer poldering technologies from the

Netherlands to treating Bangladeshis "like guinea pigs in a laboratory" (Warner 2008, 142). Others highlighted the inadequate understanding of consultants in topography, sedimentation, and local socioeconomic and agronomic conditions (Nandy 1991; National Research Council 1971, 38; John W. Thomas 1972b). Another report indicated the cause of these failures more directly, in explaining the particular inappropriateness of these highly experimental technical designs in Khulna's uniquely dynamic landscape:

> There is considerable risk in an area like East Pakistan in dependence on mechanical and human efficiency, particularly if the decision-maker, the operator of the machinery, does not share the priorities and interests of those dependent on him. The hazards of creating a situation in which the welfare of a large number is dependent upon the precise timing and reliable operation of a human and mechanical system in an environment not conducive to this type of precise efficiency constitutes a drawback of the polder project concept. (John W. Thomas 1972c, 21)

As this indictment of the CEP indicates, the failures of these infrastructures were the result more of the experimentation in technological design than of the region's particular environmental challenges. Shapan Adnan has described this experimental technocratic approach as the "dryland" view, which is directly opposed to the "wetland" vision that characterizes the region's indigenous water regime (Adnan 2009). The dryland vision entails a complete rupture with existing agrarian water management practices, as well as a failure to recognize their advantages in relation to soil fertility, the maintenance of groundwater reserves, and the natural breeding of wild fish populations. Instead, this view sees the traditional flooding patterns of Khulna's wetlands as a threat to be eradicated through central planning and landscape engineering.

The wetland approach is characterized by a recognition of the important role of flooding in the natural ecology of the region (Lahiri-Dutt and Samanta 2013). This is reflected by the different words used in the Bengali language to connote good or bad water inundation. *Borsha*, which also refers to the monsoon season and the rains that it brings, is essential for agriculture, while *bonna* refers to floods to which people are not able to easily adapt. *Borsha* for some may be experienced as *bonna* for others; for example, the monsoon rice crop (*aman*) cannot thrive without seasonal water inundation, but this same flooding is often intolerable for urban dwellers, for whom such flooding impedes the commute on which their livelihoods depend (Shaw 1989).

While donors and consultants continued to pursue the "dryland" approach by adapting and expanding the polder system as challenges arose, resistance to these new experimental designs quickly emerged. In the final chapter of

Leedshill–De Leuw's 1968 report on the progress of the CEP construction, they briefly note that several local governmental bodies in Khulna requested that infrastructures not be built in their localities due to the problems that were already emerging. The engineers, however, pressed on, explaining that "positive benefits and economic justification [could] be shown" (296), and thus construction would continue. On several occasions, local farmers took resistance into their own hands by physically dismantling sections of the embankments to restore the flow of water (Koch 1991; Leedshill–De Leuw Engineers 1968; Nandy 1991; Atiur Rahman 1995).[23] These events recall earlier resistance to embankments during the colonial period, during which administrators complained of cultivators intentionally breaching protective embankments in order to allow water to flow in to irrigate their fields, thus undermining the integrity of land reclamation infrastructures (Harrison 1875, 5; Westland 1874, 121). In the late 1980s, development agencies proposed a dramatic expansion of major flood control infrastructures built on the same technical approach as the CEP and known as the Flood Action Plan (FAP). Subsequent civil society activism against the plan was so robust that it had to be withdrawn (Adnan et al. 1992; Boyce 1990; Lewis 2010), although several components of it ended up being implemented under the guise of other development projects.

One of the most significant problems that emerged in the years following the initial construction of the polders in the 1960s, which continues to plague the coastal region today, has been the failure to plan or make provisions for their maintenance. Though this lacuna has been recognized repeatedly throughout this period of development planning, it remains a primary cause of the most serious failures of the embankment infrastructures. The CEP was designed following the Dutch polder model in the Netherlands, which relies fundamentally on considerable annual maintenance paid for by both the national government and the local municipalities (the Dutch spend about $600 million annually on maintenance alone for existing water control infrastructure, though this is expected to increase dramatically as new plans are developed to address the threat of climate change) (*New York Times* 2008; Stijnen et al. 2014). The initial plans for the CEP recognized that its success was contingent on the proper operation and maintenance of the polders; in lieu of making provisions for this maintenance, the plan indicated that it would be the responsibility of the farmers who were benefited by the protection (Leedshill–De Leuw Engineers 1968). However, even these early reports indicated an awareness that this would not be feasible. Leedshill–De Leuw's 1968 report on the CEP explained that "farm sizes smaller than about five acres would not have the payment capacity for project costs, except by reducing the farm family living expenditure" (279). With 78 percent of farms in Pakistan consisting of less than 5 acres in 1960 (Swadesh R. Bose 1972, 79), this plan to have individual farm households provide for the maintenance of the CEP was clearly not grounded in a realistic analysis of the local agrarian political economy (van Ellen 1991).

Additional problems included the recognition that it would be necessary to pump water out of the polders (as opposed to allowing gravity to propel drainage through the sluice gates), though the costs and installation of pumps were not included in the project design or donor plans (National Research Council 1971). Reminiscent of problems observed in the colonial period, the failure of sufficient maintenance of the embankments has exacerbated problems inherent in their design, resulting in the weakening and breaching of embankments, siltation of internal canals, and development of intractable waterlogging due to insufficient drainage (Koch 1991; Soussan 2000; Swingle et al. 1969). Building such a massive flood control infrastructure without the plans or capacity to maintain it was itself a kind of experiment undertaken by consultants, donors, and the local bureaucracy that relied on them.[24] Today, Bangladesh continues to grapple with the aftermath of these failed landscape engineering experiments of the East Pakistan period. Their legacy fundamentally shapes the region's vulnerability to climate change, both present and imagined.

Dispossession

These modes of experimentation in landscape engineering and land management ultimately facilitated extraction through the dispossession of local inhabitants. In imagining the dystopic present and future of the Sundarban region, the administrators of successive development regimes have sought to intervene in the landscape in ways that transform it to enable accumulation. The adaptation regime exhibits a much longer historical pattern of intervention based on normative conceptions of the value of this landscape, its resources, and the communities inhabiting it. In particular, questions about whether the Sundarban region is worth saving have been a recurrent theme in these attempts to govern the region. The 1875 *Bengal Embankment Manual*, which sought to codify the norms and laws around how and when to protect, reclaim, and reshape which lands, stated that "the only question that can arise in these tracts is whether the country to be protected is worth the cost of the protective works" (Harrison 1875, 29). This fundamental question has beset every subsequent development regime in the region. Meanwhile, the sense that the land to be protected is either not worth the cost of the protective works or not worth the risk of the experimental proposals for protection has consistently been a driver of dispossession.

The *Bengal Embankment Manual* also prescribes that the benefit of embankments must be "large and general" (Harrison 1875, 14). The vagueness of this dictum allowed for the normative assessment of what constitutes a "large and general" benefit to shift over time and to be applied unevenly. The shift from local toward central control of embankment infrastructures facilitated this development. While this centralization can clearly be observed in the experiences of the CEP and the FAP described previously, it had already begun in the colonial period.

Colonial administrators repeatedly used the unique biophysical features of the landscape as justification for shifting land management regimes in the Sundarbans (O'Malley 1908, 83). This often meant unsettling claims of the cultivators to self-determination in the use and management of their lands, a key practice that carried through to the present. The "temporarily" settled legal status of the Sundarban land tenures allowed the government to exert control over which lands would be reclaimed, how they would be taxed, and the kinds of infrastructures that could be used in protection and reclamation. Tenants in the Sundarbans who were subject to these laws repeatedly contested these fluctuations in land management policy, as demonstrated in legislative records concerning disputes over taxation, lease renewal, and land management (Pramodranjan Das Gupta 1935).

The administration of Sir Richard Temple, who served as lieutenant governor of Bengal from 1874 to 1877, provides an indicative case of the fluctuations in these administrative regimes. During a visit to the Sundarbans at the beginning of his tenure as lieutenant governor, Temple observed that the forests were home to an abundance of timber resources that were valuable for fuel for the growing capital of Calcutta, as well as for boat building (Temple 1882, 419). The timber was also necessary to the expanding rail network and, as Buckland notes, "an experiment was also being tried for employing the *sundri* timber in the manufacture of railway sleepers" (Buckland 1901, 613; see also Ramachandra Guha 1990). This new management of timber resource extraction was at odds with the existing policy of essentially unmitigated deforestation in service of land reclamation and the expansion of agricultural cultivation. In light of these observations, Temple somewhat abruptly shifted government forestry policy in the Sundarbans away from active land reclamation and deforestation and toward the controlled management of forest resources in order to facilitate their extraction and taxation.[25] By 1904, 78 percent of the total forested area of Khulna had been categorized as protected forest "as a means of ensuring a continuing supply of timber and other forest products" (Richards and Flint 1990, 27).

Temple's new policy in the Sundarbans coincided with a broader transition throughout British India toward government-controlled forest management (Brandis and Smythies 1876; Sivaramakrishnan 1999), which culminated in the

Indian Forest Act of 1878. These shifts, Ramachandra Guha has observed, were "concerned above all with removing the existing ambiguity about the 'absolute proprietory right of the state,'" and the "usurpation of rights of ownership by the colonial state which had little precedent in precolonial history" (1990, 67). The prior categorization of forested lands as "waste" was invoked in service of British claims on the unalienable "right of conquest" of the colonial government, which one British administrator described as such: "We hear a great deal about the rights of *the people*, but, as a matter of fact, the State since it assumed the administration of our forests, never has admitted the existence of any class of public rights to them. To do so would be to cease to administer them" (Amery 1876, 28). This reflects a particularly significant moment in forest management regimes in the region through the claims of the state to the absolute rights to the governance and the extraction of forest resources, accompanied by the complete rejection of any rights of the local inhabitants to the forests whatsoever. In the Sundarban estates, which were outside the Permanent Settlement, this meant that the government could levy higher rates of rent and impose new conditions on land management (such as requiring the building or maintenance of embankments) (Pramojandran Das Gupta 1935). By stemming the expansion of the agricultural frontier, this new forestry policy put increasing pressure on the zamindars, whose profits were contingent on a gradual expansion of cultivation in newly reclaimed lands. As the zamindars sought to increase their rental earnings without expanding the area of their tenure, they exerted increased pressure on the cultivators, the results of which were magnified as the effort was passed down through the hierarchy of subinfeudation (Gokul Chandra Das 1996; Iqbal 2010). As Iqbal explains, "Agrarian decline occurred as a consequence of the erosion of cultivators' entitlement to ecological resources" (187). Thus, dispossession of the cultivators in the Sundarbans was the result of both a shifting regime of forest management, as well as an agrarian class structure that had become increasingly stratified through the period of colonization.

Yet it was not Temple's object simply to expand forest conservation in Bengal. Rather, he imagined colonial forestry as a science of shifting geographies based on the requirements of accumulation at different points in time. In one of many memoirs that he wrote after his term of service, Temple argued that

> National benefit would arise if the people were to migrate from one centre of industry to another, according to need. But this would be an undertaking contrary to their disposition, and certainly beyond the power of any government. Though some classes are migratory, yet the people in the main are domestic and home-abiding. They are attached to their ancestral rights in land, are fond of the fields they till, and cling to the

humblest of their homestead.... If, however, the people were stirred by the colonizing impulse which moves hardier and sturdier races, there is still, within the bounds of India itself, a vast quantity of arable land awaiting the invasion of the plough.... [In some places,] when enquiry is specially turned towards the cultivable waste, outlying lands are found, some here and some there, the grand total of which would be anticipated by few except statisticians or surveyors. (Temple 1880, 82–83)

Temple goes on to describe the vast "wastelands" of relatively newly colonized British Burma, which, in their proximity to Bengal, offered the potential for the immigration of farm labor to expand the cultivated land within the empire.[26] Thus, even as Temple laments that moving people around to serve the needs of the state would be "beyond the power of any government," he nevertheless shaped forestry policy toward precisely that end. The dispossession of agriculturalists in the Sundarbans served to expand the state's accumulation through the extraction and taxation of resources, as well as bolstering the growth of the industrial city of Calcutta through supporting the extraction of timber for fuel and other urban requirements. The imagined role of migration in this development regime shares important parallels with the imagined role of climate migration in the adaptation regime (this is explored further in chapter 2).

These tensions between agricultural cultivation and forest resource extraction in the coastal zone were the subject of ongoing discussion within colonial administrations (Richards and Flint 1990). One particular debate concerning the formulation and application of these laws in Kolkata in 1915 illustrates these discussions and the resistance to the dispossession that they revealed. From the early colonial period to the late eighteenth century, the reclamation and settlement of the Sundarbans had been realized through the construction of small embankments around plots of land in the tidal range, in order to protect fluvial accretions. The work of constructing these embankments was carried out by raiyat laborers, while it was supported and promoted by zamindar landlords, who were benefited by the resulting expansion of the size of their landholdings. As the goals of the colonial administration shifted, however, the government's objectives came into conflict with those of the zamindars.

A recognition of the risks of premature land reclamation led administrators to attempt to better control the construction of embankments. Whereas over a century of reclamation work had relied on intensive intervention that artificially reshaped the landscape, suddenly administrators were concerned with understanding how the natural tidal patterns sustained the interests of the state and of capital. British forester E. A. Smythies observed that "the khals [tidal channels] are the natural highways for the extraction of forest produce, which is all brought out by boat, the tidal currents being utilized for conveying the boats in one direction or the other" (Smythies 1925, 41). Given this recognition of the importance of the natural flow of these waterways, it followed that the construction of embankments that would interfere with tidal flows should be controlled administratively. Thus, in 1914 a bill was introduced to the Bengal Legislative Council proposing that all embankment construction should be at the discretion exclusively of the government and impugning the "cultivators" of the Sundarbans for causing the deterioration of the tidal channels through the unauthorized construction of embankments and the obstruction of natural drainage (IOR 1915). While examining the faults of cultivators, these discussions failed to address the role of the colonial property regime in encouraging zamindars to promote this deforestation and reclamation in the first place. The papers documenting the deliberations over this bill reveal a heated debate over the state control of land and water management in the Sundarbans.

Of the five-member select committee tasked with developing the legislation, two members were Indian, Hrishikesh Laha and B. Chakravarti, the secretaries of the British Indian Association and the Bengal Landholders' Association, respectively, both of which were essentially organizations composed of and advocating for the interests of Bengali zamindars. Laha and Chakravarti objected emphatically to the proposed legislation, as well as to the characterization of the activities of cultivators in the Sundarbans that it addressed. They argued that cultivation in the Sundarbans was only possible through preventing the saltwater from coming in and also retaining rainwater for irrigation purposes. These objectives were only possible, they explained, through the local and small-scale construction and management of embankments at the cultivator's own discretion (including the liberty to cut the embankments open in order to discharge excessive accumulation of water), allowing cultivators to respond quickly and efficiently to dynamic changes in the tidal environment. The proposed legislation, they argued, would prevent the capacity of cultivators to do so and would also undermine the legal rights of Sundarban residents and cultivators to the fluvial accretions that might emerge adjacent to their lands (expanding their potential area for cultivation). At any rate, they asserted, the legislation was based on a "misapprehension of facts" on the part of the government about the natural tidal patterns of the Sundarbans and the effects of small embankments within them. Siltation of the canals, they explained, was not caused by these small embankments; it was only a naturally occurring phenomenon in these waterways that are "generally not navigable by large steamers." Laha and Chakravarti's point here is that the pattern of siltation of the waterways did not pose a problem for navigation by the small country boats traditionally used by residents both for domestic and commercial purposes; only did so for the large seafaring ships used by British traders. Beatson Bell, the member in charge of the committee, argued in defense of the bill, retorting that "It is extraordinarily difficult to define the word 'navigable'" and that "a navigable river is a river which is navigable." Bell was arguing that any channel through which any kind of watercraft, including a simple raft, could be floated was by definition "navigable," and therefore subject to government administration, concluding that "now, it is well known that rafts can float in almost every creek of the Sundarbans" (IOR 1915). The crux of the disagreement is that the Indian members representing the zamindars perceived the value of the Sundarbans for agricultural production and settlement, and thus advocated for the autonomy of the cultivators, while the British members of the committee sought a shift toward the more active state management of all the waterways and shipping routes (which would facilitate the expansion of large-scale colonial extraction from British India).

Throughout the recorded proceedings, Laha and Chakravarti explain their objections to a pattern of government interference and dispossession that the proposed bill would only entrench. They appeal to historical experience in the region, including several failed "experiments" in water and land management on the part of the government and a general pattern of "cumbrous and dilatory" government control. The British members dismiss these charges of dispossession with reference to an alternative understanding of the region's biophysical dynamics—how the land and watercourses shifted, what the effects would be of planned management, the appropriate role of human intervention and habitation. They do so through repeated appeals to scientific expertise and their interest in "proper and expedient" management measures. In the end, having been outnumbered, Laha and Chakravarti were overruled; the bill passed in 1915.

While the legacy of these debates over embankment infrastructure continues to reverberate in the landscape today, the impacts of wasteland settlement policies on land tenure have had perhaps an even more profound influence on the region's social life. Land tenure in the Sundarbans during the colonial period was shaped both by the particularities of policies put in place to support land reclamation and by the unique biophysical features of the landscape and the associated challenges of settlement. This land tenure was characterized by an extraordinary degree of subinfeudation (relative to the areas settled under the Permanent Settlement outside the Sundarban tracts), with tenancies being multiply subdivided into an extensive hierarchy of subtenures.

This exceptional degree of subinfeudation can be attributed directly to government efforts to settle new lands in order to increase tax revenues (Hunter 1875a, 62). The "mania" among government officials to survey and sell wasteland tenancies took place with so little oversight that they often usurped lands that were already occupied (Buckland 1901, 543). Leaseholders were granted large plots in the Sundarbans under "exceedingly liberal" terms that were revised four times between 1784 and 1853 (Iqbal 2010, 27; Westland 1874). These terms stipulated that no rent would be required for periods of up to twenty years; however, these leases would be terminated if tenants failed to bring significant portions of the land under cultivation within a stipulated amount of time. Because this reclamation could only be undertaken through considerable expense, the tenancies were usually taken by wealthy absentee landlords based in Kolkata, about half of whom were British (Hamilton 1820; Hunter 1875a, 1875b; Sarkar 2010).[27] Clearing the dense jungle to facilitate cultivation was strenuous, as it was carried out by hand with machetes and axes by teams of ten to fifteen manual laborers accompanied by a fakir (a Sufi holy man) entrusted with protecting them from tigers (Fawcus 1927, 31). This manual labor was necessarily undertaken annually for several years before the threat of reversion to jungle subsided, and one observer noted that it took five years before the soil was domesticated enough for the use of a plow and ten years for the harrow (32).

This formidable enterprise of taming the land, as well as the natural fragmentation of large plots by rivers and canals, made the task of reclaiming large tenancies unwieldy in the expedited time frames stipulated by the lease terms (Sugata Bose 1999, 43). Thus, reclaimed plots were substantially and increasingly subdivided at an unparalleled rate (1986, 17). This also eased the burden of the significant costs of reclamation on any single tenant (1993). This subinfeudation through a network of middlemen produced what one British settlement officer called "the most amazing caricature of an ordered system of land tenure in the world" (cited in Eaton 1993, 221), with as many as twenty subtenures parceled out for a single plot (Gokul Chandra Das 1996, 58). Land legislation explicitly allowed these middlemen to proliferate between the leaseholder and the cultivator (Field 1885; Tanigushi 1981), while the proprietors "freely exercised the power of alienation," meaning that landlords and middlemen could allocate leases to different tenants at will (O'Malley 1908, 145).

These historical modes of agrarian dispossession continue to impact rural life in this region today, contributing to the region's acute inequality of land tenure and high rates of landlessness. This inequality, as refracted through the failures of postcolonial land reform and new modes of market integration, has resulted in structural continuities that sustain patterns of dispossession, appropriation, and unequal agrarian class structure (Adnan 1999; Boyce 1987). Today, Khulna continues to have one of the most unequal rates of land distribution in rural Bangladesh (Ministry of Water Resources 2001). This differentiation also deeply shaped the social and physical geography of the region and contributed to the vulnerability of the inhabitants, as explored in more detail later in this chapter. This vulnerability in turn has shaped dystopic imaginaries of Khulna as a space of climate crisis in the present.

Throughout the subsequent history of the region, these continuities can be traced through foreign aid relations and development interventions. In an indictment of the CEP and other projects implemented through EPWAPDA, one US-AID consultant wrote, "The record of these large projects supports the conclusion that the rural people are not their primary beneficiaries" (John W. Thomas 1972b, 7). His reflection confirms that even as development regimes changed in the Delta, the pattern of dispossession continued to play an important role in shaping the landscape and the communities that inhabit it. This is further demonstrated through the expression of interest in developing shrimp aquaculture in early documents on the CEP and related development interventions. While it is commonly thought that the promotion of shrimp began in the 1980s under structural adjustment programs (Adnan 2013; Paprocki and Cons 2014), earlier development planning documents suggest that these aspirations existed from the very inception of the CEP. The stated goal of the CEP was to keep saltwater out of the polders; yet Leedshill–De Leuw's 1968 report repeatedly mentions the potential for shrimp cultivation within the polders by the use of sluice gates to bring water in and the embankments to prevent the water from leaving (Leedshill–De Leuw Engineers 1968). This potential for facilitating saline water intrusion to promote shrimp aquaculture was echoed by World Bank reports in 1970 and 1972 (International Bank for Reconstruction and Development 1970, 1972a) and pursued directly in 1969 through a USAID-supported project undertaken by researchers from Auburn University in Alabama (Swingle et al. 1969).

In a keynote lecture given at a workshop on Integrated Flood Management in Dhaka in 2014, Ainun Nishat, one of Bangladesh's leading experts in water management and climate change adaptation, drew direct links between donor visions for expanded shrimp aquaculture and inappropriate water management technologies. Weaving between English and Bengali while speaking to a room primarily full of Bangladeshi water engineers, Nishat admonished foreign development professionals who promote particular interventions without a full understanding of their relationship with the local physical geography. Another workshop on "Revitalizing the Ganges Coastal Zone" had taken place at the same conference center for the three days prior, examining similar topics but organized by the CGIAR research consortium (formerly the Consultative Group for International Agricultural Research). This conference was attended by a far greater proportion of foreigners, who gave presentations about research they had conducted and interventions they were promoting in water management and production systems in the country. In this speech the following day, Nishat reproached the experts at this prior conference for failing to understand the specificity of Bangladesh's coastal landscape and for seeking to apply foreign expertise to local problems. "They don't have a clue about what is the coastal zone of Bangladesh!" he protested, arguing that these experts had embraced an approach based on nontidal hydraulics, which was completely inappropriate to local conditions. Nishat asserted that foreigners don't understand that Bangladesh has three distinct cropping seasons, which is important to understanding the different kinds of water management techniques suited to each of these seasons. "Bangladesh is different from Malaysia!" he emphasized, obliquely referencing the findings garnered by many CGIAR experts in Southeast Asia that are frequently applied to Bangladesh.[28] The CGIAR consortium, he said, "had a huge conference[;] they paid lots of money for lots of studies, but their approach is all wrong. They want to keep the water there for shrimp." However, the conflicts between shrimp and rice were only increasing, he explained, and effectively addressing the water management challenges facing the southwestern region would require addressing these conflicts directly. Not doing so would only result in further dispossession.

As development agencies have consistently described shrimp aquaculture as beneficial, or at least benign and inevitable, they have actively ignored indications to the contrary. An evaluation conducted in 1996 of Dutch development programs in Bangladesh since independence highlights this pattern. The evaluation examined the Delta Development Project (DDP), a development program carried out in Polder 22 between 1980 and 1992 with the objective of achieving the "integrated development of land, water and human resources" (Netherlands Ministry of Foreign Affairs 1996, 142). Under the direction of a collective of young, idealistic activists who had been hired as consultants, the DDP worked closely with Nijera Kori to catalyze the successful organization and empowerment of landless peasant groups in the polder. The program facilitated access to common lands by landless agricultural collectives around the perimeter of the entire island, and in turn these collectives took responsibility for routine maintenance of the embankment. However, the report indicates that the Dutch aid agency decided to discontinue the project in the late 1980s because it was at odds with the rapid expansion and "severe pressure" toward transition to shrimp cultivation in surrounding polders (142). This recognition and failure to address the role of shrimp in agrarian dispossession in Khulna must be understood in the context of the much longer history of intervention in this region.

Human intervention in the Sundarbans region has shaped the area that is now southwestern Bangladesh and the social structures of the communities that inhabit it in profound and enduring ways. Through evolving regimes of land and water governance, including land reclamation, land tenure policy, and the construction of embankments at different moments in history, the dynamics of imagination, experimentation, and dispossession have manifested historically and continue to shape the present. Across this history, we find continuities in both modes of representing the region and, relatedly, political economies of extraction to serve the needs of a distant metropolis. Representations of the Sundarban region as risky and existentially vulnerable paved the way for interventions intended to stabilize, manage, and improve the landscape in order to facilitate this extraction.

The methods by which the Southwest was constituted as a site of imperial power, through representation and intervention, laid the groundwork for the contemporary dynamics of the adaptation regime. Any systemic understanding of the vulnerability of the region to climate change cannot fail to take into account how these historical patterns have shaped Khulna's geography as it exists today. Instead of thinking about these dynamics in isolation, we would be best served to interrogate them dynamically, and thus to understand how climate vulnerability itself has been historically produced.

2

THREATENING DYSTOPIAS
Development and Adaptation Regimes

On the sidelines of the 2015 United Nations Climate Conference in Paris, while delegates of the UN's member states were reaching a global agreement on reducing and responding to climate change, representatives of the world's leading development agencies were meeting nearby at the Development and Climate Days workshop.[1] The goal of the workshop was to discuss strategies to "seize the opportunities presented by climate-compatible development." Through lectures, panels, role-playing games, and other interactive sessions, participants discussed with an almost breathless enthusiasm the opportunities offered by climate change for realizing a particular vision of development. This vision, organizers explained, would entail "tough talk" on the transitions in energy, land use, and human habitation that they described as "crucial" and "necessary." While speakers saw these transitions as imperative due to the effects of climate change, they also saw them as "opportunities." The excitement surrounding these opportunities was illuminated by colorful neon stage lights, which bounced off the historical wooden beams of the handsomely renovated event space.

Throughout the two-day workshop, which involved eighteen plenary and breakout sessions, speakers implored the over two hundred participants (primarily policymakers, scientists, and development practitioners) to "speak the language of business." Business, we were told, is a natural ally of development and climate change adaptation. "Do you accept that in the long term, development is about deep structural transformation of economies?" the leader of one major aid agency boomed animatedly into his microphone during one plenary session. Nearly everyone in the room raised their hands in agreement. He went on:

> We need to talk about development and climate change together.... Development needs to be at the center of the conversation about climate change.... If we go ahead 30 years down the road, if we're looking at a village today, maybe no one in that village will live there anymore, and they'll all be working in a garment factory down the road. So our job [as development practitioners] is to help manage that structural transformation for the benefit of the people who live in those villages.

The crowd was energized. The speaker had crystallized the vision for the future of the gathered development agencies, one articulated repeatedly through their discourse and activities concerning climate change adaptation. The specter of climate-induced ecological crisis was translated by the speaker into possibilities for industrial growth and export-led economic development (along with the demographic shifts that will accompany them).

As I watched from the back of the room, having just arrived in Paris after two years of fieldwork in Bangladesh, I was struck by how this vision of rural futures mirrored the narratives I had heard repeatedly from development practitioners in Dhaka, Bangladesh's capital. It seemed that the village in this man's parable might easily be one of the villages in southwestern Bangladesh, in the district of Khulna, where I had worked for several years. His narrative of rural decline, as well as his normative vision of the need for urban development alternatives, aligned closely with the ways in which development practitioners and policymakers in Bangladesh discuss the future of these villages and their inhabitants. In this coastal region of Khulna, a complex of ecological and political-economic shifts threatens rural livelihoods and even the existence of rural populations and the landscapes they inhabit. In what follows, I argue that the convergence of Bangladesh's contemporary development regime with new discourses and practices of climate change adaptation is not only transforming Bangladesh's coastal geography but also shaping it as a laboratory for this type of development throughout the rest of the world. I examine the dynamics of the adaptation regime in the present (imagination, experimentation, and dispossession), and through an investigation of the groundwork laid for them through prior development regimes in Bangladesh.

This chapter examines the emergence of the adaptation regime as a mode of governing both people and landscapes in Khulna. This governance is contingent on new imaginaries of an uncertain but devastating future under climate change, as well as a discourse about the inevitability of this future both globally and in Bangladesh in particular. In response to the belief in this future, many development agencies have begun to propose dramatic (even previously unthinkable) social and spatial reorganizations of the rural coastal zone of Bangladesh, a dynamic that I have elsewhere referred to as "anticipatory ruination" (Paprocki 2019). While the vision of these agencies is predicated on many of the same assumptions and goals that have characterized the development project (McMichael 2004) since the 1950s (Hart 2001; McMichael 2008; Watts 2009), the discourse of climate change makes them different in their appeals to urgency, pronouncements of inevitability, and evocations of scientific authority. Through the adaptation regime, the dispossession of rural communities and growth of an urban industrial economy come to be seen as both inexorable and propitious futures.

I take the dynamics of agrarian change in Khulna as a lens through which to investigate the adaptation regime in its concrete manifestations. Significantly, many of the development strategies promoted under this regime have been drivers of agrarian dispossession entirely independent of their use as climate change adaptation strategies. I move back and forth between sites and scales in order to understand the interrelations between concrete strategies for intervention that drive this dispossession and the broader discourses and development imaginaries that have come to motivate and facilitate it through the adaptation regime.

Who and Where Is the Adaptation Regime?

Who are the actors who build, shape, and participate in the adaptation regime? What are their interests? What do the everyday practices of this work look like? What kinds of fissures exist in their narratives and approaches? The adaptation regime is not unified or coherent; it is not a "thing." Yet it does involve systems of practices and ideas that must be understood collectively. The adaptation regime is illuminated through both the "common sense" and the "counter-discourse" of the actors who operate within it (Abu-Lughod 2000; Wolford 2010). For Gramsci, this "common sense" is the toolkit through which people uncritically comprehend events, ideas, and processes; it is these practices of cultural hegemony through which the political-ideological status quo is perpetuated (Gramsci 1971). My ethnographic practice among these actors involved a variety of both formal and informal exchanges, from observing their public dialogues at large conferences to private conversations in offices, chats over lunch or tea, or conversations on boats and in cars in Dhaka's infamous traffic jams. Among these different settings, it became clear that these diverse actors variously advanced, consented to, and challenged public narratives of their own agencies and the broader regime of which they were part. While I would frequently hear them clearly articulate a dominant discourse of adaptation imperatives, often when I questioned these actors (at least in private), they would also cite discrepancies or acknowledge their own reservations. It is through such contestations and conjunctures that hegemony itself is actually produced (Goldman 2005, 24).

For actors within development agencies, this has often meant an explicit or implicit acknowledgment of the silences their work engendered. While there was broad commitment to the idea of the need for action related to climate change, these actors often acknowledged to me that their interventions would be the same regardless of climate imperatives. It also meant recognizing the normative work done to frame both narratives and interventions. Moreover, they were often aware of the politics that they felt they necessarily ignored. For example, I regularly asked questions of staff at large aid agencies about land politics and their programs' disregard of them in imagining climate change adaptation. I found that they frequently agreed with me that addressing land inequality could have a significant impact on the vulnerability of coastal inhabitants, but they explained that doing so would be "messy," would take more time than they had within their project cycles, and was generally beyond the scope of what they felt they could accomplish. One donor from the United Kingdom's Department for International Development (DFID), in responding to my question about concerns with land grabbing and related water management conflicts, explained, "Dealing with all that social stuff is very complicated, time consuming, and does not consume a lot of money. But if you've got a lot of money to spend, infrastructure is the way to do it."[2] The implication here is that investing in projects like embankments, which are considered less politically contentious, is less complicated for donors and, for whom large capital expenditures often become a proxy for impact of their work.[3] Yet this donor also recognized the implications of these gaps to their programming, and later admitted to me, "We could do better in understanding the political economy and governance issues before we launch into proposing solutions." My questions about development agencies' support of shrimp aquaculture were frequently met with similar agreement. While USAID is one of the major supporters of the expansion of shrimp aquaculture in Bangladesh, I heard reports from several staff there about contentious debate within USAID itself about whether the agency should support it, citing negative social and environmental impacts.

Similar contestation exists among Bangladeshi participants in the adaptation regime. Government bureaucrats and local staff of NGOs would frequently explain to me that a particular narrative or approach was necessary for the sake of the Bangladeshi nation, which needs the funding that is made available through bi- and multilateral funds to address climate change. Sometimes they would legitimize certain actions by explaining that it was the donors who made the decisions about the kinds of development activities to pursue and promote. Support for shrimp aquaculture as a climate change adaptation strategy in particular was often framed by these actors' own class positions, personal interests, and experiences. Given the investments in aquaculture among urban elites, I met many actors who had (or had family members with) personal financial interests in shrimp aquaculture; at the same time, several such people now living in Dhaka grew up in villages in Khulna and expressed conflicted feelings about the changes they had observed in the landscape through the transition from rice to shrimp. Thus, even as I deploy categories such as development practitioners, government bureaucrats, and scientists, I am conscious of their own recognition of the incoherence of their narratives along with their responsibility to advance them.

While there is significant diversity within these categories of actors participating in the adaptation regime, the categories of "local" and "international" are equally incoherent. Neither do these categories map clearly onto the interests or motivations of various actors. Both Bangladeshis and foreigners participate in the adaptation regime. Yet the experience of most of these actors is profoundly transnational, shaped by often decades of working and negotiating with a variety of international actors, both in Dhaka and abroad. Even as claims to local knowledge and perspectives are based on true experiences, the experience of being a Bangladeshi is not unitary, and is shaped by class, gender, livelihood, and a variety of other socioeconomic factors. The ideas of these actors about possible or desirable futures under climate change are shaped by these transnational circuits of knowledge and capital. Thus, even as we examine the particular historical and geographic specificity of the adaptation regime in Bangladesh, we must necessarily understand it as a node in an international political economy of development that is always and already shaped by global actors and processes.

Bangladesh as "Ground Zero"

In its short history, Bangladesh has been a key site in the global development project, a geography of imagination and experimentation with new frontiers in what Gillian Hart calls "'big D' Development," a project of intervention in the "third world" emerging during the Cold War (2001). It is in this context that the emergence of the adaptation regime in Bangladesh must be examined, both to understand the role Bangladesh has played in its emergence, as well as to better understand the regime itself. The role of Bangladesh at the forefront of climate change adaptation highlights the deep imbrication of the development project in the emergence of the adaptation regime.

The role of Bangladesh as ground zero for climate change adaptation grew out of its status as ground zero for development and colonization. This centering of

Bangladesh as ground zero was contingent on two dynamics: first, enframing Bangladesh as a perpetual "basket case," and second, establishing it as a laboratory for development practice and research. These processes continue to shape the work of development in Bangladesh and are the foundation of the adaptation regime.

Development Regimes in Bangladesh

The history of Bangladesh itself has progressed alongside the growth of global developmental imaginaries and regimes. In the aftermath of the brutal war of independence from Pakistan in 1971, unprecedented amounts of foreign aid poured into the country to finance reconstruction; by 1979 these aid flows were equivalent to 20 percent of Bangladesh's gross national product (Hartmann and Boyce 1983, 268). Foreign aid has not come without conditions, however. As Sobhan argues, it produced a belief among major donors that "the size and importance of their contribution to Bangladesh's development effort [gave] them a right to dictate how it should conduct its development affairs" (Sobhan 1982, 146). The country's first administration, led by Sheikh Mujibur Rahman, had an uneasy relationship with foreign donors, owing largely to the latter's Cold War antipathy to the new government's avowed socialism (Lewis 2011; Sobhan 1982). Shortly after gaining independence, Bangladesh was infamously referred to by an American diplomat as a "basket case," a characterization that has haunted the country ever since.[4] After Sheikh Mujib's assassination in 1975, these tensions cooled (even if the external cynicism directed toward the country did not) as the military government of General Ziaur Rahman worked closely with the World Bank (in its role as the leader of a consortium of major foreign donors) to implement far-reaching market-led reforms, including trade liberalization, denationalization of the jute and textile industries, devaluation, monetary stabilization, the establishment of early export processing zones, and the reduction of subsidies by raising prices of public goods (Muhammad 2006; Sobhan 1982; Uddin 2005; Van Schendel 2009). The World Bank (along with other foreign donors) continues to exert a strong influence over Bangladesh's national budget and policy making (Byron 2015). The results of these reforms have been extraordinary; in 2014 the Pew Research Center called Bangladesh the second most market-friendly country in the world, while *Forbes* magazine dubbed it a "capitalist haven" (*Dhaka Tribune* 2014b). These narratives reflect the overwhelming success of these capitalist visions in shaping developmentalist imaginaries in Bangladesh.

Many scholars have noted that Bangladesh's capitulation to these reforms reflected not only a transformation in economic policy and resource flows but also a surrender of considerable sovereignty over both domestic and international policy in exchange for ongoing aid commitments (Lewis 2011; Muhammad 2006). This capitulation has been bolstered by an expanding urban elite whose class interests have become bound up with the interests of foreign donors (Sobhan 1982, 200). As Lewis explained, "The content and representation of Bangladesh's economy and society had now become absorbed within the international project of developmentalism.... [Today, aid] remains a powerful influence at the level of ideas and policy" (2011, 39). Thus, early skepticism (if not contempt) of Bangladesh's right to self-determination was quickly succeeded by the hegemony of a development regime firmly rooted in the nascent neoliberal development model.[5] The significance of this developmentalism extended beyond the Bangladeshi state, as the country became a global "test case of development," in the words of two former World Bank economists (Faaland and Parkinson 1976).

This hegemony coincided with the emergence of one of the most robust local development industries in the world (Lewis 2011). The model of neoliberal development was propagated largely through an NGO sector that grew rapidly in its scope and geographic reach (Devine 2003); today NGOs have a presence in at least 90 percent of Bangladeshi villages (Siddiquee and Faroqi 2016). As these NGOs increasingly took responsibility for social welfare activities, such as education, health care, and the supply of drinking water, the provision of entitlements that were formerly considered the purview of the state was increasingly privatized (Feldman 1997; Sarah White 1999). The rise of Bangladesh's NGO sector was linked with the country's expansive capitalist market reforms. Collectively, they have provided fertile ground for the emergence of the adaptation regime.

Despite threats to state sovereignty that accompanied this period of structural adjustment, the Bangladeshi state and the national political elite who govern it continued to play an important role in the country's development regime (Ludden 2006). Hossain describes Bangladesh's regime as being contingent on simultaneous agreements between the governing national elite with both the Bangladeshi public (to ensure basic subsistence and provide protection from catastrophic flooding and cyclones) and also the international development community (which provides aid, resources, and recognition). This agreement with the latter in turn required Bangladeshi elites to open the country up to a range of experimental policy and development interventions, in effect making it a global development laboratory (Naomi Hossain 2017). At both the local and national scale, this settlement has entrenched a political system defined by patronage relations. In the last decade, however, a democratic system characterized by competitive clientelism has shifted toward a more authoritarian single-party state that has consolidated power from the local to the national scale (Lewis and Abul Hossain 2019). These multiple and shifting systems of political alliance and patronage play an important role in legitimizing the work of the adaptation regime.

From Basket Case to Development Laboratory

The role of Bangladesh in the adaptation regime is only the most recent phase in a long pattern of experimentation in social and ecological engineering. Though such experiments existed during the British colonial period, the first official use of Bangladesh as a development "laboratory" was through the Pakistan Academy for Rural Development (PARD, which was reborn after independence as the Bangladesh Academy for Rural Development [BARD]) (Ali 2019). In 1959, PARD was founded through support from USAID, the Ford Foundation, and Ayub Khan's military regime. Its centerpiece was a "social laboratory" established in Kothwali Thana,[6] an administrative unit of about 150,000 inhabitants in the southeastern part of the country (Wahidul Haque 1977). PARD produced a rural development paradigm that came to be known as the Comilla Model, and which was celebrated as a model to be replicated in rural communities throughout the "Third World" (Ruttan 1997). By 1977, the model had been used to implement rural development programs in 200 of Bangladesh's then 434 *thanas* (Azizur Rahman Khan 1979). BARD functioned both as a training facility for rural administrators in new technologies in agricultural development as well as a living laboratory for experiments in "cooperative capitalism," involving the dissemination and intensification of green revolution technologies through multiclass cooperative organizations, the provision of credit, the provision of reproductive health care for population control, and other experiments in social engineering, as well as a variety of public works, including the construction of irrigation and drainage canals and of flood protection embankments (Anisuzzaman et al. 1986).

PARD's lionized founder, Akhter Hameed Khan, was an early champion of the contemporary tradition of participatory development. A disillusioned former colonial civil servant, Khan envisioned the Comilla Model as an opportunity for experimentation in community-led development (Lewis 2019). Yet the Comilla Model did not address itself directly to the problem of inequitable land tenure and agrarian production relations (Wahidul Haque 1977). Embedded within an agrarian political economy that continued to exhibit an essentially feudal character, the benefits of the model consequently accrued to larger landholders, mostly due to elite capture and disenfranchisement of the landless (Azizur Rahman Khan 1979). The program itself ultimately exacerbated rural inequality and dispossession, as land transfer from small to larger farmers increased within the cooperatives (Blair 1978; Wahidul Haque 1977). In 1974, BARD's erstwhile figurehead, Akhter Hameed Khan (who had departed in 1971 when East Pakistan became Bangladesh), wrote, "No more shall the people of Comilla be harassed by my antics, nor I be overwhelmed by their problems.... I call myself an expert in failure" (1974, 5).[7]

Yet despite its failures in reality, the Comilla Model was largely regarded as a success in global development circles (Ali 2019; Blair 1978). This sense of the model's success can be attributed primarily to its ideological function in serving as a global platform for experimentation with new and mobile interventions conducted by a growing development industry. PARD offered development practitioners and the scientists who worked with them a platform, not only for experimenting with a variety of development technologies, but also for exploring the possibility of a model for a suite of technologies that could be universally replicable. As one social scientist involved with PARD explained, "The contribution of these studies to sociology is that they utilize concepts and hypotheses that were developed in the United States in a new setting and test the universality of the existing findings. Their peculiar accomplishment is that they do this under severely adverse research conditions" (Choldin 1969, 490). East Pakistan (and later Bangladesh) provided the perfect platform for such a laboratory because of its severe impoverishment as well as supposed lack of governing capacity, which suggested its need for external intervention. Choldin's study highlights the importance of PARD in demonstrating, not only the utility of green revolution technologies, but also their significance in establishing "social laboratories" for conducting experiments in rural development (1969).

PARD's concern with agricultural technologies converged with a large-scale promotion of contraceptive technologies. Ali writes that over the decade of its existence through the 1960s, PARD distributed "more than a million condoms, more than half a million foaming [contraceptive] tablets, and inserted IUDs [intrauterine devices] into 4,500 women" (Ali 2019). This emphasis on family planning expanded beyond the academy, such that in the 1980s, Bangladesh was at the front line of population control efforts, as neo-Malthusian discourses of resource scarcity and overpopulation framed a new generation of development programs. With support from the Ford Foundation, USAID, the World Bank, and other bi- and multilateral donors, clinical trials and other experiments with the Norplant contraceptive implant, IUDs, and tubal ligation were carried out under the auspices of development programs (Hardee, Balogh, and Villinski 1997; Hartmann 1995). In 1981, Bangladesh received the largest total and largest per capita amount of funding for population control of any country in the world, almost ten times more per capita than neighboring India (Herz 1984, 19). While these programs facilitated access to some reproductive healthcare for poor women, they were also plagued by reports of failure to obtain consent from trial participants and forced sterilization (Hardee, Balogh, and Villinski 1997; Hartmann 1995).[8]

By the early 2000s, Bangladesh was again in the spotlight of global development imaginaries thanks to its burgeoning microcredit industry. Microcredit programs, and the development and research agencies that promoted them, promised an "end to poverty" through small loans to poor rural women (Yunus 1999). Over 60 percent of rural households are members of microfinance agencies, which by 2008 claimed some 10 million members and an annual loan disbursement of US$1.8 billion (Khandker, Koolwal, and Badruddoza 2013). The proliferation of microcredit programs brought Bangladesh worldwide attention as a global model for this new development "panacea" when Mohammad Yunus (and the Grameen Bank, which he founded) won the 2006 Nobel Peace Prize. Critics of microcredit, however, pointed out that despite unrestrained enthusiasm (even "evangelism") for this new model for rural development, little evidence existed to suggest that it had any real impact on the reduction of poverty (Duflo et al. 2013; Rogaly 1996). Moreover, several studies have found that microcredit is implicated in the exacerbation of indebtedness, social and cultural alienation of women, and other forms of rural dispossession (Karim 2011; Paprocki 2016).

Imagination

The adaptation regime shapes Bangladesh's contemporary development landscape and imaginaries of what is possible in the time of climate change. Yet equally, through the adaptation regime, Bangladesh is situated in the center of a global imaginary of climate crisis and adaptation. In interviews I conducted as well as public events in Dhaka, development practitioners and government officials alike asserted the inseparability of climate change from any possible imagination of Bangladesh's future. Every conversation about Bangladesh's development over the next decade to the next century thus must reflect on and respond to the possibility of climate crisis, which is a continuously asserted existential risk. The notion of this inseparability is an important tenet of the adaptation regime and is what makes Bangladesh the ideal site for its establishment.

At a public seminar in Dhaka in January 2015, the secretary of Bangladesh's Ministry of Environment and Forests, the primary ministry tasked with managing efforts at adapting to climate change, appealed to the audience of adaptation experts (practitioners and academics) to recognize the importance of Bangladesh in this global adaptation landscape. "This is the ground zero of vulnerability," he proclaimed. "[It is] disaster's homeland.... We are living testimony of what is happening due to climate change." The secretary's concern with framing this relationship indicates the importance of establishing Bangladesh both epistemically and ontologically in this global regime. The international finance and support of adaptation agendas in Bangladesh are contingent on the ideological consensus concerning the nation's vulnerability. It is this sense of crisis that creates opportunities for new frontiers of development and accumulation (Swyngedouw 2013b). This recognition of Bangladesh as "disaster's homeland" both facilitates the acquisition of resources for and catalyzes the transformation of rural spaces into laboratories of adaptation.

Memoirs of "The End of the World"

Much of the work of the adaptation regime, then, involves imagining what the future will look like, and that often has a dystopian quality (Swyngedouw 2010). These dystopic imaginaries reflect both the anticipation of future climate crisis as well as anxieties related to contemporary agrarian livelihoods. In this latter sense, these imagined dystopic conditions predate and are also extrinsic to the impacts of climate change. Khulna is the perfect place to carry out this work of imagination because many researchers and development practitioners already regard it as a sort of dystopia. "Munshiganj is the end of the world," said an American consultant hired by USAID to lead CREL (Climate-Resilient Ecosystems and Livelihoods), their flagship adaptation program in southwestern Bangladesh. At the time of this conversation, I had just returned to Dhaka from a visit to the union of Munshiganj, home to several of CREL's "model villages." I interpreted this reference to the end of the world as a comment on the remoteness of this area. Munshiganj is the southernmost union of Khulna Division, which is considered widely to be the most vulnerable region of the world's most vulnerable country. The Southwest is cut off from Dhaka and the rest of Bangladesh by the Padma River; traveling there takes the better part of a day and involves a variety of different modes of transportation. Munshiganj in particular is where the road ends, bordered to the south by the Sundarban mangrove forest and to the west by India.

However, besides conveying remoteness, this statement reflects a deeper and often-expressed anxiety about the uncertain, risky, and dystopian future of the Southwest. The fact that much of this landscape is already experiencing ecological crises facilitates a vision of Khulna as climate dystopia, and the sense that it may even already be upon us. Climate change experts link this region to the adaptation regime by circulating time-lapse maps of its coastline being inundated by sea level rise. Visiting researchers and consultants make day-long field visits to see settlements precariously perched on embankments. They accompany their accounts of these visits with photographs of erosion and postcyclone cleanup efforts. These narratives offer a prophetic slippage between the present and future tenses of this climate dystopia, auguring the climate crisis that will come, or that may have already arrived. The imagination of Bangladesh's dystopic future has become "common sense" by drawing on this imagery of the coastal region today. Yet this ambiguity about whether Khulna is experiencing a present-day dystopia or will experience it in the future allows for a spurious insinuation: that the challenges faced by coastal communities today are the direct or even exclusive result of climate change.

The texts produced by these field visits are semiotically rich, and both shape and are shaped by the way the region is understood in relation to climate change. One donor quipped to me that "there is not a single document in this country that does not start with 'Bangladesh is the most vulnerable country in the world to climate change.'"[9] His comment reflected not only the awareness of this sense of vulnerability, but also awareness of its hyperproliferation. In addition to the vast body of grey literature, visiting interns, journalists, consultants, and others working in the adaptation industry produce a body of literature we might call *climate crisis memoir*, which appears in blogs, posts on NGO websites, the local and international English-language press, and undergraduate and graduate theses.[10] In these accounts, authors detail stories of desperate people whose homes they have visited, and who have most likely been displaced by some event that the author links to climate change.[11] Stark photographs depict signs of ecological change, such as erosion, cracked earth, barren landscapes, absent explanations of local political ecologies or broader context.

Among the more evocative climate crisis memoirs produced about Bangladesh are two episodes of an Emmy-award winning American television show called *Years of Living Dangerously*. In the show, celebrity hosts travel around the world looking at the impacts of climate change and the scientists who are confronting it. At the end of the first season, Michael C. Hall, star of the popular television series *Dexter*, travels to Bangladesh, visiting slums in Dhaka as well as villages in Khulna and talking with some of the country's foremost leaders in climate change and adaptation planning. Through these experiences, he explores climate change as a driver of a series of demographic, economic, and ecological concerns. He starts the first episode by giving us a tour of the *Dexter* film set, while he explains in a voiceover:

> Inside the air conditioned, air brushed world of Hollywood, climate change isn't something you have to think about too much, unless you really want to. But I know I am living in a bubble, and I know there are places outside my bubble where climate change is impossible to ignore. At the top of that list is Bangladesh, where I've heard that climate change is a matter of life and death.

After spending a couple days in Dhaka, he heads for Khulna, while describing southern Bangladesh as "the front line in this country's battle against climate change." Traveling around Khulna by boat and seaplane, with ominous orchestral music in the background, Hall tells us, "I've heard people call Bangladesh a land of rivers. But that didn't prepare me for what I'm seeing. There's water. *Everywhere*." He is particularly intent on learning about climate migration, which he tells us is causing Dhaka's growing slums to swell. He has a handful of conversations with people who have been displaced within rural communities, asking them questions about how cyclones and climate change have impacted their lives and whether they think they'll be able to stay in Khulna. After visiting a family whose home had been severely damaged by river erosion and cyclonic winds, he reflects that "while these people may not want to leave, soon they may not have a choice.... Seeing these remnants of people's homes makes it hard to grasp why they try to stay. Especially because all of this will eventually be under water."

Hall's interviews are punctuated by foreboding depictions of the deltaic landscape, implying links between what we see in the video images and climate change itself. At one point we are following behind him on a ferry, watching him look unnerved as he brushes his teeth, staring out at the river, while he tells us in a voiceover, "As I make my way through southern Bangladesh, what I'm seeing is hard to take in. Half-submerged trees. Houses built on stilts to keep them above the rising water. There's something apocalyptic and other-worldly about it all." One might respond that the trees of Hall's dystopia are mangroves (half-submerged by their very nature) and building houses on stilts is a common feature of the region's vernacular architecture (Rashid and Ara 2015). These concerns would seem to be beside the point. Rather, Hall's depictions articulate a rich imagination of a dystopic future that the people of Khulna are already experiencing in the present. Their context beyond global climate change is immaterial to the narrative. The larger stakes of these threats become more apparent later when Hall interviews the then-US ambassador to Bangladesh, Dan Mozena, with whom he discusses the serious security threats posed by the climate refugees who will (and, in this frame, already are) pouring out of this region. Hall explains, "Southeast Asia [*sic*] is already a politically volatile region. To understand how climate change could create even more conflict, just look at a map."[11] Ambassador Mozena tells Hall that "a failed Bangladesh, a Bangladesh that has not adapted to the impact of climate change, a Bangladesh that implodes—that's a tremendous security risk to the region, and to *us*." Then the screen cuts abruptly to *New York Times* columnist Thomas Friedman interviewing Barack Obama about how the impacts of climate change in "poorer countries" are one of the greatest security threats facing the United States. The narrative arc of Hall's climate crisis memoir thus comes full circle in rendering America itself as the subject of these threats of climate change.

In another memoir, entitled "The Unfolding Tragedy of Climate Change in Bangladesh," published as a blog on the Scientific American website, Robert Glennon, a University of Arizona law professor describes a visit to Bangladesh in 2016,

December 9, 2016. "How do they survive?" I kept wondering as I walked the alleys of Old Dhaka, the capital of Bangladesh, a country with a population of 164 million on a landmass the size of New York State. People seem to be everywhere in Dhaka, in a churning frenzy of rickshaws, CNGs (Compressed Natural Gas Vehicles), taxis, buses, horse-drawn carriages and people—16 million and rapidly growing. The newest arrivals, mostly climate change refugees, end up in decrepit slums.

December 18, 2016. "What will the sea do next," I thought when I visited the remote village of Premasia, Bangladesh, at the junction of the Sangu River and the Bay of Bengal, south of Chittagong. The schoolchildren greeted us with spontaneous joyfulness, full of hope, despite the visible aftermath of Cyclone Roanu, which struck in May 2016, washing away homes and permanently ruining croplands from salt deposits. Their three-story concrete school, raised on stilts, served as a cyclone shelter during the storm. Isolated palm trees, now surrounded by water and beach, are haunting reminders that here once stood someone's home. Rising sea levels are turning land into sea bottom, driving some people farther inland. Others rebuild repeatedly, just as Sisyphus kept pushing the rock up the hill. (2017)

Glennon's account of his visit to Bangladesh reflects a particularly common literary device in these texts, in which the memoirist recounts their own speculative apprehension about the survival of the communities they have visited. This speculation reflects both a disquiet about the present living conditions of the community as well as anticipation of crisis in an uncertain future.

Other memoirs highlight these anxieties about the coastal landscape alongside descriptions of the lives of the people who inhabit it. A sense that the people of the Southwest are merely being "kept alive," in the words of one donor, pervades conversations about the development of the region. "These people are doomed"; their lives are "shit," in the words of others. "You get this grim feeling that they have no future," explained one researcher to me, about her visit to Khulna, "You just think 'you guys are fucked.'" These comments convey a sense not only of future threats, but also the notion that Khulna is already a kind of dystopia. What they elide is their own assumptions about what exactly it is that makes the lives of Khulna's inhabitants so "shitty" (a word I heard repeatedly in this context throughout my fieldwork). While the threat of rising waters is usually at the forefront of these kinds of reflections, they often blend into more nuanced descriptions of the experts' own imaginations of the challenges of rural livelihoods—that it is not only climate change, but also the difficulties of the agrarian livelihood in Bangladesh generally that make these people's lives undesirable. The "backbreaking" work, in the words of one official from DFID, of being a farmer in the remote, hot, and crowded swamp that these people call home is a cause of great concern for many development practitioners and other visitors. Their comments articulate a broader assumption that climate change adaptation experts working in Bangladesh express repeatedly. That is, the objects of their adaptation programs are people who have no hope and are living on the brink. They are people who are in need of alternative pathways out of their current lives and livelihood conditions, and these are pathways that development agencies are uniquely positioned to provide. The realities of climate change are in many ways incidental to this imagination of desperation and need for development.

Fundamental to these perceptions of agricultural livelihoods is a linked assumption that farmers do not want to continue being farmers and that climate change adaptation therefore offers them a welcome opportunity to move out of agrarian livelihoods. At a public lecture entitled "Addressing the Climate Challenge in Asia: Role of Finance ++," Dr. Bindu Lohani, the Asian Development Bank's (ADB) vice president for knowledge management and sustainable development, addressed this assumption directly in response to a question about how the ADB thinks about population and land issues. He said, "Who is going to do farming? The sons of farmers don't want to do it. They have the same aspirations as you and I! Agriculture will have to be looked at totally differently. In the future, we'll look at the farmers instead as the CEO of the farm." Indeed, Lohani's assumption that farmers have the same aspirations as a room full of donors and development practitioners is foundational to the discourses of the adaptation regime. Testimonies from farmers in Khulna presented in chapters 5 and 6 controvert these assumptions, instead indicating continued aspirations to agrarian livelihoods.

These memoirs construct climate misery as an object of development, serving to justify development interventions in the name of adaptation. They operate as memoirs not only in the sense of biography, but also in the sense that they memorialize an anticipated loss of life, an anticipation that is an artifact of their own design. The memoirist is the subject of the narrative, with the residents of the coastal region serving as their object. What is most troubling about the narratives is their incongruence with the stories of the residents themselves about the complex historical and contemporary dynamics shaping their communities. Narratives of climate crisis often suggest, either directly or obliquely, that residents have failed to understand the changes reshaping their landscapes. One memoirist wrote, "Climate change is the buzzword of the decade, and yet the very people who live on the coasts of Bangladesh, directly impacted by global warming, rarely understand the term" (Khanom 2016). On the contrary, my own ethnographic research suggests that not only can most residents of these communities supply a clear scientific description of climate change and its global geopolitics, they can also offer a detailed, nuanced perspective on its articulation alongside a range of other dynamics that have shaped the region.

This focus on the imagination of climate crisis among development practitioners, policymakers, journalists, and others demonstrates the preoccupations they bring to their work in planning for a climate-changed future. It matters how these people think because these preoccupations shape their understandings of what is possible and desirable, and in turn these understandings motivate the concrete actions and interventions with which they shape the landscape and the communities that inhabit it.

Experimentation

Besides imagination, the adaptation regime operates through material practices of intervention that actively reshape landscapes and communities. Much of the enthusiasm among adaptation experts in Bangladesh is centered on the successful transformation of the coastal zone into a "laboratory" in which innumerable experiments can be carried out to test what adaptation to climate change might look like (see Hennessy 2013; Knorr-Cetina 1992; Tilley 2011). One expert explained that Bangladesh is "the place where the rest of the world comes to learn how to tackle climate change." Consultants, planners, and researchers celebrate the development of Bangladesh as a landscape of "innovation," where the very fact of destruction creates opportunities for experimentation with new ideas and technologies. As the idea of Bangladesh as "adaptation laboratory" is developed and celebrated by foreign and Bangladeshi adaptation experts alike, it becomes clear that this success has less to do with the promise of any particular intervention or set of interventions than it does with forging a landscape of experimentation. These interventions are thus considered successful as experiments even when they participate in the production of crisis.

To catalogue potential adaptation experiments, NGOs and research organizations have begun to compile "inventories" and "checklists" that list the wide range of technical interventions that they have identified as possible responses to climate change that are available for replication in climate-vulnerable communities around the world.[13] Inventories are documented in reports and spreadsheets that are circulated among various agencies and presented in seminars in Dhaka. Some examples include an (unpublished) "Adaptation Technologies Matrix" developed by the Asia Pacific Adaptation Network, a table of "Adaptation Measures" published in the USAID report, "Adapting to Coastal Climate Change" (USAID 2009), and a "Climate Change Adaptation Inventory" developed by the "DEltas, Vulnerability and Climate Change: Migration and Adaptation" (DECCMA) project, which involves a consortium of researchers from Bangladesh, India, Ghana, and the United Kingdom. The DECCMA inventory, which is among the most robust iterations of such tools, contains 122 "documented examples of observed adaptation" from Bangladesh, India, and Ghana, including "any choices or adjustments to climate variability and change. These adjustments may be in response to, or in anticipation of, real or perceived climate stressors" (Tompkins et al. 2017, 5). At a "Dissemination Workshop" for this inventory held in Dhaka in 2015, researchers explained that they had identified possible adaptation options for the inventory using keyword searches for both academic and grey literature in Google, Google Scholar, Academia.edu, and other academic databases. In this way, potential adaptation strategies come to be understood tautologically as any actions that someone has already called adaptation strategies. At the workshop, researchers noted that they had confronted an analysis problem in that some adaptation options are considered successful by some people but unsuccessful by others. All these adaptation options made their way into the inventory, regardless of this interpretive analysis. What the inventory also misses are any ways in which people navigate their changing environment that are not referred to as adaptation strategies.

In an exemplary demonstration of such inventories, the NGO WorldFish created a "Climate Smart House" for a single family in one coastal village. Raised up on concrete stilts, the house is stocked with technical fixes to match every climate-induced problem WorldFish staff could imagine, from the "sanitary" latrine on the roof to the rain-fed fish tank underneath (E. Hossain, Nabi, and Kaminksi 2015). When I visited the Climate Smart House, its residents generously gave me a tour of its many features, most of which were in various stages of disrepair. One WorldFish staff member whom a colleague and I interviewed in Dhaka in December 2014 noted, however, that "it's not for community replication—it's for the donors," and that it existed now principally "for the website." I interpreted this to mean that the power of the Climate Smart House was more ideological than material, to the extent that it served as a demonstration of possible modes of experimentation, ideally to garner additional funds for future projects to be implemented by WorldFish itself. It is in this epistemic sense that the Smart House serves the adaptation regime.[14]

Some NGOs are developing adaptation "technology parks," where assemblages of possible interventions are collected and modeled (e.g., Siddique 2015). In describing one such park (and inviting me to visit), a European consultant responsible for developing the project explained in June 2015, "We have the space to play around, and to invite other organizations to play around with us." High-tech experiments like "geosynthetics" (polymer sheets used to stabilize eroding coastlines) and "ultra-violet disinfection" (used to purify drinking water) are positioned neatly alongside more systemic interventions such as coastal zoning and saline aquaculture expansion. Adaptation becomes common sense through this proliferation of interventions, and the selection, appropriateness, and geographic targets of interventions begin to appear self-evident. An adaptation expert at one UN agency explained to me in March 2015, when I asked her about the scope of their work on climate change, "We don't define adaptation, we just implement adaptation projects." It is through the adaptation regime, then, that the interventions that can be considered "adaptation projects" are determined and adaptation is rendered technical (Li 2007).

These geographies of experimentation are managed through the spatial governance of interventions by various development agencies. One World Bank consultant shared with me a map of the coastal region that he said was replicated in almost every internal report or proposal circulated within and among development agencies conducting adaptation work in the coastal zone. The map depicts a color-coded diagram of all of the fifty-seven polders of the southwestern region highlighted in various neon shades, with a key indicating which polders had been claimed by which development projects and which were "available" for new proposed experimental interventions. The map recalls those produced at the Berlin Conference during the Scramble for Africa, as do frequent comments by representatives of various agencies referring to their project sites in the possessive case (e.g., "that's one of *our* polders").

The threat and discourse of the dystopic future of the Southwest becomes both a rationale for experimentation and an excuse for its failures. The spatial imaginary of a landscape that is already on the verge of annihilation allows planners to treat the Southwest as an adaptation tabula rasa. It also erases histories of intervention in the region that have shaped the contemporary ecological crisis (in particular, commercial shrimp aquaculture, which is now proposed as a solution to the very crisis to which it contributed). One donor discussed this approach as a policy of "no regrets," suggesting that if the landscape is going to be destroyed anyway or perhaps is not "worth" saving, then there can be no regrets in conducting experiments with uncertain and potentially destructive results. When it comes to the expansion of shrimp aquaculture, practitioners contend that any production in a landscape that they deem to be on the verge of collapse is a success of adaptation (in the absence of any alternative possibility for comparison). If shrimp aquaculture takes the place of rice agriculture (and the livelihoods and communities that are dependent on it), then the idea that the latter is not viable—or will not be viable in the near future—reframes this dispossession as a fortuitous bonus.

One manifestation of this landscape of experimentation is the constant production and dissemination among development agencies of information through "evaluations" of NGO development projects, a body of knowledge production that exists in its own methodological and epistemic plane (see also Ferguson 1994).

When it comes to adaptation to climate change, these evaluations often serve to demonstrate simply that an experiment was conducted rather than that any particular result may have been garnered. Adaptation options are produced for the sake of "demonstration," a category that indicates an experiment sitting outside any particular social context. The problem is that many such experiments are exactly just that—experiments. As an example, the administrator of one UN agency described to me his frustration with the discrepancy between a small "test plot" with a signboard in English and a technology that "actually works" in the field and that farmers are adopting. Throughout rural Bangladesh, seemingly every possible space, from drinking wells and convenience stores to many agriculture and aquaculture fields and fertilizer factories, is dotted with such colorful signboards emblazoned with conspicuous logos indicating the NGOs that have implemented and the donors that supported the project. That these signboards are frequently printed in English, a language unlikely to be read fluently by a single resident of any given Bangladeshi village, indicates their function as symbols for donors on site visits (or for pictures for promotional websites, as one NGO staff member pointed out to me). However, this administrator explained to me that particularly in the case of agricultural adaptation experiments, the results of these test plots are often much better than they are if and when farmers implement the techniques in their own fields. "We're promoting or pushing technologies without really understanding what's going on," he remarked to me. It is precisely, then, the role that such projects perform in shaping the geography of experimentation that makes them adaptation strategies.

Dispossession

The discourse of Khulna's dystopic future often begins and ends with a question, which was articulated succinctly by one World Bank administrator: "Is it even worth keeping people there?" Of course, the answer to this question is deeply normative. Who gets to ask a question like this, and who gets to answer it? And what happens if the answer to that question is negative? These speculative calculations and the results of responding to them reflect a process of dispossession in the adaptation regime.

Garments, Shrimp, and Dispossession

Since the 1980s, the political economy of development in Bangladesh has been characterized by export-led growth dominated by the country's garment sector. This development vision has required dramatic and systemic social and economic transformation throughout the country. For the World Bank, pursuing this vision requires a concerted transition of Bangladesh's labor force into the nonfarm sectors (Muzzini and Aparicio 2013; *Daily Star* 2014c; World Bank 2014a).[15] The World Bank's 2013 "World Development Report" singled out Bangladesh's export garment sector for contributing to an urbanization rate of 30 percent, double the rate in 1980 when the garment sector was still in its infancy (World Bank 2012, 197). The sector is now the second largest in clothing exports in the world (after China), growing from an annual export value of US$300 million in 1983 to US$32 billion in 2018 (WTO 2019, 120), and accounting for more than 84 percent of all exports from the country in 2019 (Paul 2020).

Yet the growth of the frozen shrimp export industry is also central to this development vision, both in its contribution to export diversification as well as its role in transforming the agrarian economy. This vision of export-led growth sees shrimp as contributing to Bangladesh's economic autonomy, as reflected in a slogan outside the Fisheries Department compound in Khulna, which can be translated as "an autonomous Bangladesh cultivates more fish." With shrimp exports valued at US$550 million in 2014, it is the fastest growing agricultural sector, with an average expansion of 6.2 percent annually between 2011 and 2016 (Ovi 2014; World Bank 2016). It is seen as critical to the expansion of "non-crop agriculture," which the World Bank regards as a more productive sector than crop agriculture, and therefore necessary to the growth of Bangladesh's economy (World Bank 2016). Between 1984 and 2015, the area under shrimp and prawn cultivation in Bangladesh grew from 64,246 hectares (ha) to 275,274 ha (Belton 2016; Pokrant 2014).

Relatedly, and perhaps more importantly, the "productivity" gains garnered through the growth of shrimp aquaculture have precipitated significant rural-urban migration (Adnan 1993; Belton 2016; Datta 2006; Pokrant 2014). Though circular seasonal labor migration has a long history in this region, the rise of shrimp has driven more permanent rural out-migrations over the last several decades (Paprocki and Cons 2014). This availability of a seemingly endless supply of cheap migrant labor from rural areas has contributed to the vigorous growth of Bangladesh's garment industry (Siddiqi 2000). As one World Bank report explains, "Improving rural productivity by modernizing agriculture and diversifying nonfarm activities, in order to free up manpower for use in more productive activities, is also essential for growth" (Muzzini and Aparicio 2013, 48). The transition from rice agriculture to shrimp aquaculture has motivated a loss of agricultural livelihood opportunities, contributing to this process of "free[ing] up manpower." Precise figures are unavailable due to the diversified and seasonal nature of agrarian livelihoods, though respondents whom I interviewed estimated that shrimp cultivation requires somewhere between 90 to 99 percent less village labor than rice cultivation (these numbers are explored further in chapter 5).

FIGURE 2.1. Shrimp *ghers* in Khulna.

Photo by the author.

Moreover, this transition also accounts for the loss of other economic and subsistence activities, such as poultry and livestock rearing, native fish culture in homestead ponds, homestead fruit and vegetable cultivation, and the gathering of cooking fuels (Datta 2006). These losses are largely due to rising soil salinity resulting from long-term shrimp aquaculture, with the salt from *ghers* (shrimp ponds) steadily encroaching into homesteads, making even the land left for small garden plots largely infertile (see figure 2.1).[16] In this context, the question cited previously as to whether it is "worth keeping people" in villages in Khulna is inflected not only by the awareness of the certain and uncertain threats of climate change, but also by an existing political economy of development that is driving dramatic social, ecological, and demographic transitions. Indeed, shrimp aquaculture drives agrarian dispossession whether it is promoted as a strategy for adaptation to climate change or otherwise.

Promoting Migration

It is in this context that the discourse of migration as a strategy for climate change adaptation has emerged. Through the adaptation regime, climate migration has been embraced as an opportunity for a particular vision of development in the coastal zone and the country in general. This emerging vision of migration as adaptation is not unique to Bangladesh (Farbotko 2010b; Felli and Castree 2012; Tacoli 2009). In linking visions of uninhabitable rural spaces with those of the economic opportunities offered by migration, this narrative proliferates both through discourses surrounding climate refugees as well as in particular development strategies carried out in the name of adaptation (Hartmann 2010). This narrative is embedded with the fundamental assumption that people living in coastal communities in the Southwest do not want to remain where they are and feel that migration is desirable.

Many donors have developed programs that aim to promote this idea directly. At one event in April 2014 at a large conference hall in Dhaka, approximately fifty representatives of major donors, NGOs, and research and government agencies gathered to discuss the advantages of such an approach. Each attendee was greeted at the door with a complimentary coffee mug bearing the words, in bright green and red letters: "LIVELIHOOD MIGRATION: Not a threat, A tool for climate change adaptation." During the seminar, a film was screened featuring interviews with men who had moved from the Southwest to Dhaka and were working as rickshaw pullers. One of the men interviewed pled to the camera (in Bengali), "I pray to God that I am able to go back to my village and farm again." That this statement contradicted the upbeat message the organizers were trying to convey seemed to be lost on most of the workshop's attendees, who responded with a discussion not of how to help these dispossessed farmers return to their homes, but instead focused on celebrating this vision of the national economic benefits of migration. An official from DFID, in commenting after the film ended, discussed the need for greater recognition among donors and government agencies of the "Climate Change Migration Dividend." By this, he explained, he meant the benefits to national development of creating a workforce of people who have migrated out of climate change–affected areas and into urban areas where they can participate in the industrial, export-oriented economy.[17] This understanding of the development potential of climate migration is embraced by the Bangladeshi government in addition to the donor community; the 2009 Bangladesh Climate Change Strategy and Action Plan, citing the impending displacement of 20 million people in the "near future," explained that "migration must be considered as a valid option for the country. Preparations in the meantime will be made to convert this population into trained and useful citizens for any country" (MoEF 2009, 17). The implication by the government (here diverging from that of most donors) is that countries of the Global North must accept migrants from Bangladesh who are threatened by the impacts of climate change.

This tension between donors and local proponents of migration as adaptation is significant, though it has received surprisingly little attention in local discussions. While there is consensus among these local and foreign adaptation experts about the need for rural out-migration as an adaptation strategy, there is not agreement about where these migrants will go. One DFID expert described this to me in a private interview as the "UK red line," explaining that "what we can't handle is 100,000 Bangladeshis showing up on the shores of the UK because of the loss and damage argument." By "loss and damage" he referred to the emerging but contested principle in the UNFCCC international negotiations about the responsibilities of countries in the Global North (which are responsible for the bulk of historical carbon emissions) to compensate countries of the Global South for losses incurred as a result of climate change, including accepting climate migrants. Conversely, while local adaptation experts often discuss opportunities for development in urban migrant-receiving communities, they often suggest obliquely (but sometimes overtly) that these migrations should also be promoted abroad—to the Middle East in particular, as well as the Global North for wealthier migrants.

This Climate Change Migration Dividend theory reflects the growing discourse within the adaptation regime asserting that agrarian dispossession is both inevitable and desirable. An executive at another large international donor agency that was funding local programming for climate change adaptation explained to me during one of his brief visits to Bangladesh the potential he saw in the contemporary moment in Bangladesh to shift away from the logic of rural development, meaning improvement of rural livelihoods in situ. He contrasted this view with a more recent movement that he saw as grounded in the acceptance of the risks of climate change, coupled with a recognition of a broader fundamental and inevitable rural-to-urban economic transition. He explained that his own interpretation of the need for climate migration had more to do with the desires of rural inhabitants to relocate because they would prefer an urban livelihood. He explained,

> There are a lot of people moving because it [life in rural communities] is absolute shit and they want to get out of it. If you look at it economy-wide, and sort of, you've got to stand back and look at the demographic transition that's occurring in any country, I don't mean the population, I mean the transition of the economy from a rural to an urban one is something that's happening and will go on happening.

This comment reflects the synthesis of an awareness of the impacts of climate change with a normative perception of the value of agrarian livelihoods in a rapidly transforming economy and ecology.

The donor continued by expanding on the role of experts in promoting this transition away from an agrarian economy, "I do think that when you're working on climate change, it's about trying to introduce that vision of what the future will look like." Here he explains the broader function of adaptation experts in securing the hegemony of the adaptation regime. The regime itself is contingent on the articulation of Khulna as a space without a future. To that end, many donors have developed programs to support off-farm employment and urban development as key components of their funding portfolios for climate change adaptation. Some examples of programs cited by donors and NGOs for generating off-farm employment include training in rickshaw repair, garment manufacturing, and shrimp value-chain work (such as shrimp net building and assembly line processing). One researcher described this work as part of a broader vision to develop "alternative mega-cities" in Bangladesh to which migrants can transition, explaining that "Khulna has the potential to become a huge mega city" because of its port and the Special Economic Zone in nearby Mongla. The expansion of major export processing zones in periurban areas throughout the country is seen as a necessary step in this planned urbanization.

While public discussions about the use of migration as an adaptation strategy focus on the benefits of these urban and periurban transitions, in private many donors speak more openly about what they consider the necessary dispossessions in rural communities that will affect such migrations. For example, a representative of USAID's largest adaptation program explained that the approach of their work is to get people to move away from their rural communities (as opposed to supporting sustainable development within them). Another donor representative discussed the ways in which their "resilience" approach was fundamentally at odds with a rights-based approach, which insists that people have a right to stay in their homes, describing this alternative approach to resilience as a "brutal" but necessary logic. To that end, he asked, "Why are we going on investing in these places without a hard-nosed analysis of whether these places are worth saving?" This, then, is the explicit articulation of the implicit assumption of the necessity of dispossession to the adaptation regime. That is, rural livelihoods will play a diminished role in the future of development under climate change.

Indeed, many practitioners focus directly on the role of these migration patterns in promoting development in Bangladesh. One member of a panel on climate migration at a conference on climate change in Dhaka in 2015 described development policies that help people stay in their communities of origin as a "policy disaster." She continued,

> Voluntary migration of some members of the family should be used as a tool for climate change adaptation. If we are too romantic in thinking

about helping people to stay in their places of origin, then we are trapping them in chronic poverty. . . . One has to think *big*. Instead of looking at migration as a problem of urbanization, we need to recognize that development will never happen if we don't encourage migration.

As this declaration of the benefits of climate migration highlights, despite the trauma implied by dystopian imaginations of climate change in southern Bangladesh, experts I spoke with were surprisingly sanguine about the future. Even as it draws attention to the catastrophic potential of climate change, the adaptation regime proposes that the crisis of climate change should be treated as an opportunity. The notion of turning threats into opportunities has become the virtual mantra of the adaptation regime. This is a foundational logic in many proposed climate change adaptation strategies. It closely resembles what Naomi Klein calls *disaster capitalism*: "orchestrated raids on the public sphere in the wake of catastrophic events, combined with the treatment of disasters as exciting market opportunities" (2008, 6). Klein situates disaster capitalism more broadly in relation to neoliberal capitalism, in particular the intensified post-9/11, post-Katrina neoliberalism that is marked by a sense of uncertain but perpetual crisis (see also Adams 2013; Gotham 2012).

In Bangladesh, the notion that those rural livelihoods that are most vulnerable to climate disaster are already obsolete facilitates this focus on the opportunities that are opened up via climate change. Experts express (to one another, at conferences and meetings, and in newspapers and other public fora) the need to be "positive" about the potential benefits that can be derived from climate change, not to be afraid of change or of experimentation, and, indeed, to have the "courage" to do so. The discourse shifts the focus onto the positive impacts of the inevitable destruction that will take place due to climate change, thus reframing dispossession as progress.

The adaptation regime reshapes Khulna's social and physical landscape through a dialectical exchange between, on the one hand, material interventions in the landscape and the communities that inhabit it, and on the other, the epistemic construction of the limits of possibility for its future. Practitioners wield the threat of a future dystopia under climate change while at the same time responding to a contemporary rural political economy characterized by a state of development that they already regard as dystopic.

To return, then, to the parable with which we began at the Paris climate conference, development practitioners who celebrate the "deep structural transformation of economies" are the actors who perform the work of imagination, experimentation, and dispossession that constitutes the adaptation regime. The village whose inhabitants will be "working in a garment factory down the road" is the landscape on which the work of the adaptation regime is imagined and performed. Yet the geography of this adaptation regime is not restricted to these villages. Much of the work to forge the regime itself takes place far from Bangladesh's coast—in Paris, Dhaka, and elsewhere. It is through the dialectical exchange between the ongoing transformations of the coastal ecology and the epistemic rendering of what its future can and should look like that the region itself is transformed.

3

OPPORTUNITY/CRISIS

Knowledge Production and the Politics of Uncertainty

In March 2014, the *New York Times* ran a cover story with the provocative title, "As Seas Rise, Millions Cling to Borrowed Time and Dying Land" (Harris 2014). The article examines what it calls the "uncertain future" of coastal communities in Bangladesh's southwestern delta region. It weaves together scientific projections of sea level rise and climate change impacts in the region with intimate stories and photographs of villagers grappling with an increasingly volatile landscape. The author laments the "millions living on borrowed time in this vast landscape of river islands, bamboo huts, heartbreaking choices and impossible hopes." These "hopes," the author explains, are primarily pinned on the possibilities of staying in (or returning to) villages in the coastal belt that the author describes as "doomed." The article thus demonstrates a common refrain about this region and its inhabitants that circulates in popular media, academic, and development policy discourses. The uncertainty of environmental change evinces the impossibility of a future of this landscape and the communities that now call it home.

The images in the article, taken in Bangladesh's Dacope subdistrict, on an island that I have visited several times nestled alongside those where my own field sites are located, were familiar to me, but the narrative (though not surprising) was incongruous with those I myself have heard there.[1] The photographs depicted vast gray landscapes degraded through salination by commercial shrimp aquaculture, which over the past three decades has taken over this and surrounding islands through rampant, unchecked land grabbing. However, though these vast tracts of shrimp ponds were depicted prominently in the article's photographs and the effects of their cultivation were evident in the lack of vegetation and visible salt deposits, the article failed to even mention the presence of aquaculture in the area, its grave impacts on the ecology, or the deep imbrication of these concerns with those related to climate change. It is clearly far more compelling to think of Bangladesh's coastal region as a space that is "disappearing" than a space in which the social dynamics of production have been transformed through their imbrication with global export markets, thus causing social and ecological devastation.

While these apparent omissions in this piece appear stark, they are not anomalies. Indeed, the piece both contributes to and draws on a narrative about this region, the Southwest of Bangladesh, that frames a particular understanding of its ecological past, present, and future. The article is representative of the ways in which the things that are known, thought to be known, or currently being investigated about the ongoing ecological changes taking place in the Southwest are taken as scientific facts, and specifically that these "facts" suggest that the region is, in the author's words, "doomed." The article also represents the limited frame within which these ongoing ecological changes are understood and the kind of narratives that are mobilized, as well as those that are elided.

This naturalization of crisis lacking an in-depth social and political analysis is common in mainstream narratives about climate change, and should therefore perhaps not surprise us. Yet, both in its analysis and in its choice of Bangladesh as this site of inevitable crisis, the story contributes to a "common sense" understanding about climate change with effects that extend much deeper. Popular narratives about Bangladesh's climate crisis reveal this common sense, which is never autonomous from hegemonic ideologies. Not all communities are said to be "doomed" by climate change, even in discussions of serious climate impacts. As we will see later in this chapter, the circulation of these catastrophic narratives can also shape the science that is produced about ecological change in the region (and in turn the ideas about how to intervene to adapt to those changes).

In what follows, I explore the politics of knowledge that shape both popular narratives reproduced through the adaptation regime and the production of scientific knowledge about this region, along with their interconnections and the normative assumptions that inform them. These narratives circulate both in the Global North and in communities of adaptation planners and policymakers in Bangladesh, among both foreigners as well as Bangladeshis invested in the adaptation regime. I examine how uncertainty is *practiced* to shape landscapes and communities. Uncertainty is not a static condition. It is mutable and negotiable. It changes as it traverses different research programs and policy dialogues. It is often both the cause and the result of contestation. When confronted with the existential threat of ecological collapse, how do the politics of uncertainty shape decisions over where and when there can be no hope and no possibility of return?

I suggest that uncertainty can be practiced in both the presence and the absence of information. The idea of uncertainty does work when it comes to planning how to adapt to climate change. The question examined in this and the following chapter is: *What work does it do, and for whom?*

Between different processes of producing and acting on knowledge, uncertainty often emerges as the result of deliberate choices. In examining the production of knowledge about climate change in Bangladesh and associated discourses of uncertainty about the future, I find that what is referred to as uncertainty is in fact often obfuscation. This obfuscation can be intentional or it can be the result of the ideological power of common sense shaped by narratives like the one examined previously about Bangladesh's "doomed" coastal communities. Uncertainty over climate change is real. But it is also mobilized through the adaptation regime to redistribute power and resources. Uncertainty can be a context and a resource for dispossession. It can be the terrain on which resource conflicts under climate change are contested. While uncertainty can be used and also produced through continuous obfuscation, it is not itself an operative logic. What I call the politics of uncertainty, thus, is precisely a matter of which particular objects of knowledge come to be seen as uncertain and how some stories are emphasized over others to the benefit of people in positions of power.

This study of the adaptation regime is an examination of transformations in the political economy of development in a moment of acute global ecological change. In this and the following chapter, I discuss how science is being carried out and used in this period. I demonstrate that this involves two deeply interconnected contestations, related to: (a) understandings of the causes of ecological change, and (b) what the change should mean for political-economic transformation. Both these contestations are informed by an analysis of the concrete material impacts of climate change as well as a normative orientation toward thinking about how to reorganize our societies (or not) to respond to it.

So why retain this focus on "uncertainty" in a context in which much is actually known, if not said? Because uncertainty is a dominant narrative in the global governance of climate change. That which is said to be known or unknown and how that certainty or uncertainty is acted on is always a matter of power and politics. While gaps in information and understanding, especially about the future, are ubiquitous, it is also true that what is said about existing knowledge is always shaped by structures of power, from which no system of knowledge is independent. By focusing on uncertainty as an unstable and fundamentally political category, I reveal how these power structures of knowledge production are enrolled in the adaptation regime.

In this chapter, I explore how these politics and practices of uncertainty are manifested in the process of conducting research on the dynamics of ecological change in coastal Bangladesh. In the following chapter, I focus on the circulation of knowledge about ecological change through a focus on the interface of research and intervention in development and adaptation. In this chapter the focus is on research led by scientists employed by academic institutions. Yet both chapters examine the politics of uncertainty through the deep imbrication of the dynamics of knowledge production and circulation among researchers and practitioners. Throughout, I attend to the political work that the idea of "uncertainty" does, and the forms of knowledge and exercise of power that it elides. These power dynamics not only shape how we understand what is happening in coastal Bangladesh; they also have real material impacts on the lives and livelihoods of people who inhabit the region today.

Instead of arbitrating what science is "real" or what claims are most accurate, relevant, or valuable, in this chapter I examine what is broadly understood by scientists about ecological change in the Southwest. I do so in order to demonstrate how that knowledge is circulated to make particular claims about possible or desirable futures (and to diminish others). I examine not only what scientists, policymakers, and development practitioners, respectively, believe is known about ecological change and the future, but also their own perceptions of the utility of this knowledge within regimes of climate change adaptation and development.

This work builds on a robust literature in political ecology and science and technology studies investigating the social context and epistemological politics of climate science (Beck and Mahony 2018; Demeritt 2001; Günel 2016; Jasanoff 2010; Miller 2004; Nightingale 2017; O'Reilly 2017; Rice, Burke, and Heynen 2015; Swyngedouw 2013b; Vaughn 2017a). These social dynamics involve both the ways in which social relations shape the production of scientific knowledge about climate change and also how these social relations inform policy outcomes dependent on that knowledge. Beyond the social relations that govern the production of knowledge, I am concerned with the circulation of knowledge about environmental change, its drivers, and possible responses (that is, whose knowledge is valued and given weight in popular and policy discourses). This latter focus highlights how the production and circulation of knowledge are intertwined with one another. The politics of uncertainty emerge through this relationship between the two.

Understanding the Ecological Crisis in Southwestern Bangladesh

Among the broad reflections of physical scientists studying the drivers of change in the Southwest, two clear analytical themes emerge. The first is that at the local level, it is not possible to analytically separate the effects of climate change and sea level rise from other drivers of change, such as shrimp farming, large-scale engineering projects, and cross-boundary water-sharing issues—there is no way of knowing, several of these scientists have explained to me, the relative importance of any of these dynamics in driving the ecological changes most often attributed to climate change.[2] The second, relatedly, is that while it is difficult to parse these various drivers, it is clear that climate change is not the only driver of these transformations, and, as most of them insisted, it is probably not the primary driver of change.

A third, related concern that many physical scientists stressed in both public and private conversations was that the scientific data about ecological change in the Delta is often misunderstood. Many cited concerns about their own data or that of their colleagues being misquoted, misinterpreted, or oversimplified, leading to flawed narratives about the scientific evidence of present or future change. One prominent social scientist, whose work has drawn considerable attention to the challenges posed by climate change to communities in the coastal region, expressed dismay about the "stylized" way in which these impacts are discussed, particularly in popular media portrayals of climate science. He explained that these stylized narratives about the threats of climate change to Bangladesh reflect basic misinterpretations of the geomorphology of the region.

A combination of interconnected ecological problems poses threats to production and habitation of rural areas throughout the southwestern coastal zone (Nicholls et al. 2015). It is not my intention to mediate between narratives highlighting these various drivers of ecological change and strands of reasoning that exclude any of them in understanding the transformations currently taking place in the Southwest. Rather, I emphasize that each has played an important role, and I examine the dynamics through which knowledge about various drivers of change is elided.

While the historical challenges and patterns of water and land management examined in chapter 1 continue to manifest in the coastal region today, they have taken on new significance in relation to climate change. The predominant environmental concerns in the coastal region are related to patterns of tidal activity at the shifting interface of land and water. Because of the proximity of these dynamics of tidal activity to concerns related to sea level rise, there is often ambiguity or uncertainty about the precise causes of change and the boundaries between them. This also generates uncertainty about the role of human intervention at this interface of different drivers of change.

Soil salinity is an important example, as it is a constant subject of debate among researchers and development practitioners. During the wet season, from June to October, the monsoon rains gather along the rivers from upstream; as they flow into the estuary, they push the brackish tidal waters back down toward the saline

Bay of Bengal. During the dry season, from November to May, with less water flowing down through the rivers, the tidal waters push back up the rivers and creeks, bringing salinity with them. The result is a salinity frontier that moves up and down throughout the year in the coastal region (Clarke et al. 2015; Lázár et al. 2015). Crops are selected for cultivation in large part based on their tolerance to the level of salinity at a given location. Sea level rise is likely to impact this frontier by causing the tides to move further inland, thus linking soil salinity to climate change (Dasgupta, Kamal et al. 2014). However, sea level rise is not the only factor driving this shifting salinity frontier. When India built the Farakka Barrage (which began operations in 1975), much of the water from the Ganges was diverted away from Bangladesh (Kimberley Anh Thomas 2017a, 2017b).[3] While these numbers are disputed, the Bangladeshi government claims that the barrage reduced the flow of water from the Ganges to Bangladesh by 70 percent (Hossen and Wagner 2015). The withdrawal of upstream water flow resulted in increased salinity in the coastal region, as tidal waters moved inland (Bradnock and Saunders 2000; Chapman 2007; Ito 2002). Finally, as explored in chapter 2, shrimp aquaculture has been perhaps the clearest human-induced driver of increased soil salinity, as land is intentionally inundated with saline water in the dry season to flood ghers for shrimp cultivation (Tanzim Afroz and Alam 2013; Clarke et al. 2015; S. A. Haque 2006). Salinity from the rivers affects crop production when it gets into the soil, which can happen through accidental inundation (when the embankments are breached), intentional inundation (for shrimp aquaculture), or an excessive use of saline water for irrigation. Salinity is leached out of the soil annually when the land is flushed with monsoon rain (Clarke et al. 2015). This flushing process often does not take place in the context of year-round aquaculture production. However, if salinity seeps into the underground aquifer, then it is a much more serious concern. This can happen either through the continual inundation for aquaculture (Barraclough and Finger-Stich 1996) or through groundwater depletion (through excessive withdrawal by deep tube wells for irrigation) (Davis and Ali 2014). Though this salinization of aquifers is already happening, it is believed that sea level rise could also contribute to saltwater intrusion into the aquifers (Clarke et al. 2015; Saleemul Huq, Ali, and Rahman 1995; World Bank 2010).

These drivers of salinity are also linked to other environmental changes in the Delta. Waterlogging, which is discussed in more detail in the following chapter, is closely related in the sense that persistent inundation with saline water drives the salinization of soil and groundwater. Sporadic and chronic waterlogging have several different causes and drivers, all of which are a source of ongoing debate. In discussions about climate change, the most important of these is sea level rise, which is consistently invoked as the greatest climate-related threat to Bangladesh's low-lying Delta. The rise of the river levels relative to land levels is a topic of considerable attention by the community of scientists working in this region (S. Brown and Nicholls 2015; Kay et al. 2015; Nicholls et al. 2013; Pethick and Orford 2013). Yet the term "sea level rise" is often used as shorthand to refer to *absolute* sea level rise, which is the increase in the absolute volume of the oceans as a result of climate change–induced factors such as the melting of polar ice caps. Absolute sea level rise is thus tied to global patterns of environmental change (Ballu et al. 2011). Relative sea level rise, however, refers to the observed difference in sea and water levels in a particular coastal landscape. Thus it is a dynamic that can only be observed locally, as it interacts with the vertical movement of coastland and a variety of local drivers of change. Nicholls and Goodbred argue that there has been an "over-emphasis on the issue of global-mean sea-level rise, versus other possible changes," and they highlight a variety of human-induced drivers of relative sea level rise in the Bengal Delta (Nicholls and Goodbred 2004, 11). In Bangladesh, land subsidence (related largely to the polder system, as explored in chapter 1) is a major contributor to relative sea level rise (Auerbach et al. 2015a; Brammer 2012; S. Brown and Nicholls 2015; Darby et al. 2015).

Land erosion is another dynamic of physical transformation that is often conflated with absolute sea level rise (A. Ahmed 2011; Brammer 2014a). As discussed in chapter 1, the Bengal Delta, being exceptionally active in its ongoing formation, is in a constant state of erosion and accretion. As observed in the *New York Times* article discussed at the beginning of the chapter, this land loss related to erosion is often attributed to sea level rise induced by climate change. Hugh Brammer, a scientist who has studied Bangladesh's physical geography and agricultural systems for over 50 years, refers to these claims conflating erosion and sea level rise as "wildly mistaken assertions" (Brammer 2014b, xvi) and highlights the historical process of constant erosion in the Bengal Delta. There is also scientific consensus, however, that relative sea level rise will exacerbate existing patterns of coastal erosion (Wong et al. 2014). Despite land loss in many parts of the Delta, Brammer's analysis suggests that between 1984 and 2007, the Meghna estuary alone experienced an annual net land gain of 19.6 square kilometers. This augments Allison's findings that between 1792 and 1998, the Delta saw an average annual net land gain of 7.0 square kilometers.[4] Building on an analysis of similar trends, Darby and colleagues projected that climate change could result in a net increase in these rates of land accretion, as the rivers may carry an increased sediment load to the Delta from the melting Himalayan glaciers (Darby et al. 2015). This would suggest that, contrary to ominous popular narratives of Bangladesh disappearing, climate change could in fact cause the country's land mass to grow larger.

Finally, increasing vulnerability to cyclonic flooding and storm surges is also considered a significant consequence of climate change in Bangladesh (Akter and Mallick 2013; Dasgupta, Huq et al. 2014), with a wide variety of drivers. The notion of "vulnerability" itself reflects a range of social, political, economic, and historical dynamics (Watts and Bohle 1993), and thus vulnerability to cyclonic activity in Bangladesh must be understood as a complex relationship between a variety of transformations. In the physical sense, however, this increased vulnerability in Bangladesh is also the result of transformations in the landscape. Specifically, the deforestation of the Sundarbans, both related to recent expansion of shrimp aquaculture and other historical patterns explored in chapter 1, has diminished the critical buffer against cyclones that the mangroves provide (Barua, Chowdhury, and Sarkar 2010; Islam and Haque 2004; Rahman and Rahman 2015).

In recent years, all of these dynamics have been attributed primarily to climate change, and, indeed, their intensity and amalgamation have amplified the sense of climate crisis. Nonetheless, they are each embedded in complex histories and political economies of the region (Sally Brown et al. 2014), histories that can be masked by narratives of the ethical neutrality and scientific credibility of this new climate science consensus. The task of recognizing or assigning relative weight to these various drivers of change is a significant source of uncertainty among scientists and development agencies alike. While most scientists recognize that each of these drivers play a role, not all development practitioners do. For example, many practitioners involved in shrimp aquaculture programs suggest that shrimp farming is a response to, rather than a cause of, salinity; the implications of this disagreement are examined further in chapter 5.

Yet even as this physical science of ecological change can reveal the silences in myopic narratives of climate change as the exclusive driver of ecological transformation, these studies of past and future change are inseparable from normative understandings of what constitutes a desirable future, particularly where it comes to studying adaptation. In reflecting on this broad spectrum of drivers of coastal environmental change, a group of leading natural scientists wrote in a commentary in *Nature Climate Change* in 2014, "Herein lies a paradox: economic and population growth can increase risk, but economic growth and prosperity promotes adaptive capacity" (Sally Brown et al. 2014, 752). This is a key point at which the production of scientific knowledge is embroiled with the adaptation regime. What kinds of prosperity promote "adaptive capacity," and for whom? Does it matter how this prosperity and adaptive capacity are distributed within and between communities? The paradox outlined by these scientists is at the heart of contestations over understandings of vulnerability to climate change in Bangladesh and potential adaptation responses. Even as they recognize that certain human interventions might transform landscapes in ways that increase their vulnerability to climate change, they also suggest that those same human interventions might increase the ability of communities to respond to those transformations. What is less commonly recognized is that the promotion of prosperity and adaptive capacity for some will come at the expense of others. Decisions about how to navigate this paradox are thus political decisions based on normative visions of the future under climate change. Uncertainty both emerges from and becomes a key instrument in navigating these contentious politics. Thus, by not addressing these questions of distribution directly, claims of uncertainty can be used to entrench inequality and vulnerability.

The Scientists

My approach to this interpretation of ecological change is motivated by interviews and participant observation with researchers studying the ecological and geomorphological impacts of climate change in southwestern Bangladesh. This work is often conducted by large collaborative interdisciplinary consortia of primarily physical scientists, sometimes also including social scientists with expertise in survey methods. Both within and beyond these consortia, this work is carried out by a wide variety of researchers of different nationalities, institutional positions, and areas of disciplinary expertise. My intention was to understand who these people are, how they do their research, what it shows, and how it engages with and impacts development policy. The diversity among and collaboration between researchers working in this field makes it difficult to neatly categorize them. I spent time with Bangladeshi and foreign researchers, university faculty, staff of NGO research centers, and consultants hired by donors and NGOs for specific projects. These people frequently occupied several of these categories simultaneously or at different points in time. In particular, the boundaries between foreign and local knowledge and expertise are not clear, as many scientists have worked and been trained both in Bangladesh and abroad.[5] Moreover, foreign researchers often rely heavily on their Bangladeshi counterparts for fundamental information about the country both within and beyond the Bangladeshi scientists' areas of expertise.

One instructive example of the range of different kinds of disciplinary expertise being applied to these questions in Bangladesh is a major consortium project called ESPA Deltas (ESPA stands for Ecosystem Services for Poverty Alleviation), which took place between 2012 and 2016. I had heard about this project repeatedly in Dhaka from several development agencies that intended to put its findings to use in the design of their adaptation programs, and I came to know it better when I met affiliated British researchers at a conference on climate change adaptation in the Netherlands. ESPA Deltas was composed of a multidisciplinary and multinational team of researchers studying environmental change and social vulnerability in the Bengal Delta. Supported by a group of British research and development agencies including the UK Department for International Development (DFID), the Economic and Social Research Council (ESRC), and the Natural Environment Research Council (NERC), the goal of the program was to support policymakers in understanding ongoing environmental change in the Delta and the role of policy in shaping it. Like much of the scientific research being carried out in this area, the program did not exclusively address itself to climate change, but instead took climate change as one of many drivers of change affecting the region that it intended to understand collectively.

The consortium of which ESPA Deltas was composed included faculty from major British research universities with expertise in coastal and environmental engineering, ecological economics, hydrology, water policy, geology, and human geography, among many other disciplines. Several of them have been lead or contributing authors to the reports of the Intergovernmental Panel on Climate Change (IPCC). These researchers partnered with a broad spectrum of Indian and Bangladeshi researchers, including several faculty members from the Institute of Water and Flood Management (IWFM) at the Bangladesh University of Engineering and Technology (BUET). IWFM is one of Bangladesh's premier institutions for expertise in water engineering, and its faculty are almost universally enlisted in studies or programs concerned with hydrology or water engineering by the Bangladeshi government, foreign aid agencies, and development practitioners alike. While its faculty have exceptional expertise and experience with hydrology and water engineering in Bangladesh, they rarely if ever have funding for implementing independent research of their own design and conception. This was true of the ESPA Deltas project, which was shaped by the funding and conception of the study in the United Kingdom, despite contributions of local partners to its design and execution.[6]

When I interviewed one Bangladeshi government water planning official about the problems with the polder system (discussed in chapter 1), he brought up IWFM and other Bangladeshi water management experts in the course of talking about the kinds of expertise that could be employed in conceiving of a strategy to address the technical problems with the design of the polders. We discussed a new World Bank project conceived of to address these problems (see also Paprocki 2019). The Terms of Reference for this project stated explicitly that only an international consulting firm could be hired to carry out this work.[7] The government official lamented to me that IWFM faculty had the knowledge and expertise to do this kind of work, but they were never given the chance and instead were always included as the subsidiary consultants to teams of researchers from outside Bangladesh. If they were given the funding and opportunity to think creatively about what a new "sustainable polder concept" could look like (the stated goal of this World Bank project), would they come up with something entirely different from a foreign firm?[8] In speculating broadly on his own question, the official suggested to me that if there could be a water-engineering strategy inspired by indigenous water management systems and the Bengal Delta's unique tidal landscape, the scientists at IWFM would be among the best people to devise it. Nevertheless, he speculated, the politics of research funding in Bangladesh are such that they would not be given this opportunity.

These unbalanced relations of collaboration reflect a broader pattern in the production of climate knowledge, which was observed by Corbera et al. (2015) in their study of authorship of citations found in the IPCC Working Group III report on mitigation. The authors found that the research cited in the report revealed persistent, unequal collaborative relationships between researchers in the Global North and Global South, with authorship dominated primarily by institutions and scientists in the Global North. Rochmyaningsih has identified similar unequal authorship patterns, highlighting the self-perpetuating nature of the links between unequal authorship and unequal research funding (2018).

In addition to the IWFM faculty, staff from other government and NGO research institutes were also participants in the ESPA Deltas consortium, including the Bangladesh Institute of Development Studies, the Center for Population, Urbanization and Climate Change at the International Centre for Diarrhoeal Disease Research of Bangladesh, and the government's Water Resources Planning Organization (WARPO). Finally, these agencies also developed partnerships in Khulna with local NGOs whose primary work involved connecting nonlocal researchers and development agencies with local communities through identifying research field sites, organizing brief visits to the region, coordinating logistical details, and even giving presentations on the social and cultural context of the area under study. While this last category of program partner is not a research institution, these partners also play an important role in shaping the researchers' understanding of the region—whom they talk to, what they see, how they see it, and the contextual details that inform their experiences. These aspects of the research design are not incidental or accidental. The information and perspectives that are chosen shape the knowledge that is produced in fundamental ways. They reflect existing forms of knowledge, power, and resource distribution, and in turn they shape whose interests are reflected in the scientific knowledge that is produced about the region. Thus, the lapses that may be observed in the kind of information that is produced may be understood as uncertainty, but this is an uncertainty that has been

FIGURE 3.1. Sundarban Shrimp Private Limited.
Photo by the author.

actively produced by relations of power that benefit some groups at the expense of others.

For example, in 2014 I joined the ESPA Deltas project team on a field visit arranged by one such local partner that involved a tour by boat into the Sundarbans. On the way, we stopped at a massive commercial shrimp aquaculture operation, which was unlike any other I had seen in Khulna. Owned by a businessman in Khulna, the Sundarban Shrimp Private Limited enterprise was spread out over 60 acres and surrounded by high fences and lampposts (see figure 3.1) We were introduced to the manager of the operation, the only worker in sight, who told us about the operation's impressive productivity and economic returns. When we reboarded the boat and were served giant Khulna tiger shrimp for lunch, I wondered aloud to some of my companions about who had lived in that village before Sundarban Shrimp Private Limited had arrived. I asked the same question again after lunch when the boat made a stop at a small settlement of landless people living on an embankment, who had apparently been displaced from their homes following a recent cyclone. They now worked collecting shrimp fry with nets in the river.

The decision not to introduce these researchers to communities where agriculture has persisted (or been reintroduced) or where residents remained to share different perspectives on shrimp aquaculture certainly shaped the way the researchers were able to see Khulna and the livelihoods of its inhabitants. The implications of this decision were demonstrated to me later in the day in a discussion about the consortium's research on out-migration and questions embedded in the overall research program about place-attachment and what makes people want to stay in undesirable locations. Of course, understanding this area as undesirable is a particular perspective evidently shaped by these relationships between research partners and interlocutors. Examining the broad spectrum of actors who are involved in the production and dissemination of knowledge thus facilitates an understanding of the power relations shaping what is known, what is thought to be known, and what remains uncertain about environmental change and life in Khulna today.

The most important thing I learned from the ESPA Deltas researchers involved the interactions between climate change and other drivers of environmental change and how these interactions are being actively shaped in the present. As one of the scientists involved in the project explained to me, even though they see themselves as people who have "built careers on" the study and dissemination of knowledge about climate change ("We're no climate change deniers!" one of them remarked), they consider it critical to recognize that there are many other social and environmental processes taking place, all of which are important to understand collectively. Most important, they explained, is that there are things that can be done to mitigate the impacts of climate change (and, indeed, other environmental transformations) and that decisions are being actively made in the present that shape the landscape now and will continue to do so in the future under climate change.

In addition to this work being done inside Bangladesh, there is also a great deal of research being done about Bangladesh by researchers outside the country, using computer-modeling methods and remote-sensing data. One leading environmental scientist who has worked in Bangladesh for several years explained that he had presented a paper based on satellite data from Khulna at an academic conference and also had visited the region repeatedly to verify the analysis and understand it better. I found that scientists who study environmental change in Bangladesh regarded this "ground truthing" component of their research methodology with varying degrees of seriousness, ranging from casual sightseeing visits to insightful ethnographic interest that shaped research questions and conclusions. I found this latter commitment to be the case with the scholar mentioned here. The paper he was presenting at this conference concerned dynamics of environmental change in the Sundarbans and revealed that he and his colleagues had found no evidence of recent forest degradation there. Following his presentation, two other scientists who had separately conducted studies of the region using different methods from those of the presenting team to analyze remote-sensing data stood up to challenge his findings, arguing that they had each found that there was extensive degradation in the Sundarbans. At least one of these scientists had never been to Bangladesh, the researcher explained to me later. He responded that they had seen no evidence of this degradation either in their satellite data or in their field visits; the dissenting scientists, whom he described as "very emotional," were incredulous with disbelief and certain that this region must be experiencing degradation.

The impression of the scientist making the presentation was that these other scholars at the conference panel had already formed assumptions about the certainty of ecological degradation in Bangladesh and that because of these assumptions, they were unable to consider the possibility of alternative findings based on different data and analysis. "How could you be so confident when you had never even been there?" the scientist asked. I asked him if he had observed this kind of dissent involving the certainty about degradation relating to other regions or any of the other field sites where he works. He responded that "it is definitely the worst in Bangladesh—and you know why, right? Because Bangladesh is *the poster child.*" By this the scientist meant to explain that Bangladesh has become a place that epitomizes these imaginaries of ecological degradation such that they shape the very science about that degradation directly. While contestations over scientific analysis are certainly nothing new, his experience and reflections reveal the role that dystopic imaginaries of climate crisis in Bangladesh can play in shaping production and analysys of scientific data about the region and how those assumptions are circulated and reproduced within and between scientific and popular discourses about environmental change in the country. This example also demonstrates that gaps in knowledge are not a static or immutable condition, and it shows how researchers' understandings of ecological change are fundamentally shaped by the preoccupations they bring to them (Watts 1983).

Circulations of Climate Science in Bangladesh

In Bangladesh, as elsewhere, conversations about climate change perpetually invoke climate science, and those engaged in them are eager to discredit denialists. Yet my research on policy making surrounding how to address climate change in Southwest Bangladesh revealed complex questions about what claims are supported by this science, and, indeed, strong (though usually discreetly articulated) concerns about the need for better or "real science" in policy making. As one donor working on climate change issues in Bangladesh explained to me:

> Of course, there's an interest in calling everything "climate change" right now, because that kind of puts you into this category of being "the most vulnerable country to climate change. . . ." But I think we're very, very weak on the science. We have a lot to do. One of the things we often say to the government is that we actually have to become better in that, because just saying that you are the most vulnerable country to climate change and that there will be 30 million climate refugees in the future and that, you know, most of the country will be gone, is probably not enough to keep the pity going for a very long time. If there are questions about what has really caused it, and where is the money going, the money can dry up very, very quickly. Which is also sad, because, you know, I think climate change *is* happening.

As this donor's apparent unease suggests, there is a gap between what is said and what is known about climate change in Bangladesh, and this gap has potentially significant implications for how climate change is addressed there.

The comments of this donor are not exceptional. Throughout my research, I frequently encountered questions among donors and policymakers about what claims are "supported by science" and how they can include better or "real" science in policy making. There is also concern among some scientists about the invocation of scientific expertise where evidence is deemed to be insufficient or excessively politicized; as one environmental scientist explained to me, "There's been a proliferation of scientists—or people who call themselves scientists. Climate change is hardly even a scientific question anymore." While this scholar would certainly not dispute that there is a great deal of critical and rigorous scientific research ongoing about the causes and impacts of climate change, his comments reflect frustration with how discourses about scientific evidence are deployed in ways that distort its findings to bestow credibility for political purposes.

One staff member working on projects for climate change adaptation at USAID in Dhaka told me that he estimated that approximately 50 percent of the funds the agency spends on climate change are related to knowledge production, knowledge management, or knowledge training and added that among his colleagues at USAID and other donor and development agencies, the production, management, and dissemination of adaptation knowledge is their primary interest in relation to climate change. The result is the amassing of a great body of data that frames the challenges and possible interventions in the Southwest in relation to climate change. Yet as the actors and relationships within ESPA Deltas demonstrate, knowledge about ecological change in the region and the appropriate interventions to address it are at once distinct and also mutually constituting. The imbrication of the two through the production of environmental knowledge alongside development interventions thus shapes how the changes are understood as well as how they are addressed. Stott and Huq (2014) confirm this in their study of Community-Based Adaptation practitioners in Bangladesh, which found that the circulation of knowledge about climate change and adaptation is both limited and textured by existing power relations within and between the global, national, and local levels.[9]

A paper published in 2016 by two researchers at the Lamont-Doherty Earth Observatory at Columbia University highlights important ways in which these circulations of dystopic imaginaries about Bangladesh might intersect with the production of scientific knowledge about climate change in the country (Chiu and Small 2016). The paper examines Cyclone Sidr of 2007, a storm that caused rampant destruction across the coastal region. Yet in examining the drivers of this destruction, the study finds that damages were likely caused by high wind speeds as opposed to flooding caused by storm surges, which has been the widely accepted explanation among scientific modelers. The paper used data collected from tide gauges managed by the Bangladesh Inland Water Transportation Authority (BIWTA) to measure the actual tidal activity at fifteen different locations throughout the coastal zone during cyclones occurring across the past four decades. It compares the observed data of storm surges (meaning the amount of sea level rise at a particular moment in time associated with atmospheric storm activity) with a series of studies that have examined these storm surges using computer-modeling methods. The paper responds to widespread reports of up to 7 meters of storm surge heights in "a considerable body of literature" on the region, which the authors note has attracted significant scientific interest due to the vulnerable nature of Bangladesh's delta region. These reports of dramatic storm surge heights garnered particular attention in the aftermath of Cyclones Sidr and Aila (in 2007 and 2009, respectively), and they have been linked with increased frequency and severity of cyclonic activity related to climate change. The authors note that "the focus of these studies is often to provide recommendations for decision makers rather than present new data on storm surge associated with cyclones" (Chiu and Small 2016, 1150), reflecting the fundamentally political and policy-oriented nature of many studies of cyclones in Bangladesh. Contrary to popular reports and these modeled results, Chiu and Small find that tidal gauge data indicate maximum storm surge heights of Sidr and Aila were approximately 2.6 meters, and that in many locations during Cyclone Sidr, the surge height actually decreased.

In the Discussion section of the paper, the authors take up the question of why their analysis of storm surges based on observed data could depart so conspicuously from previous studies that used modeled data. They propose that these prior analyses of higher surge heights of 2 to 7 meters could be based on models derived from reports of storm surges after the cyclones, rather than observed data based on in situ measurements. If Chiu and Small are correct in their analysis of their own data and its discrepancy with the findings of others, then what we see here is the existence of (and a latter challenge to) a scientific consensus based on computer models derived from reports of a particular driver of catastrophic event that may not have existed, resulting from researchers creating models to fit catastrophic popular reports instead of observed changes. This is not to deny the very serious impacts of each of these storms; indeed, together they were responsible for thousands of deaths and the damage they inflicted on the embankments and other coastal infrastructure caused incalculable harm. The *Daily Star* reported that Sidr killed around 4,000 people and affected around 1.2 million, while Aila killed 190 people and affected around 3.7 million (Farid Hasan Ahmed 2017). Yet if these impacts were the result of wind speeds as opposed to storm surges, the discrepancy between these studies is significant. It indicates how a scientific consensus can emerge around the circulation of a particular imagination of a dystopic environment, regardless of whether that imagination is borne out in reality. Understanding the true drivers of environmental change is critical to developing strategies to address it.

Producing Uncertainty about Climate Migration

The previous example suggests how discourses can circulate about dynamics of environmental change that inform both policy and the production of scientific knowledge even in the absence of clear information. Yet uncertainty can also be actively produced through obscuring information that is otherwise available. The epistemological politics of discourses surrounding "climate migration" offer one such example, in which the intentional production of uncertainty over migration patterns serves to obscure and facilitate agrarian dispossession (see also Brammer 2009; Hartmann 2010). In chapter 2, I explore in greater detail the normative orientation of development practitioners and donors toward the promotion of climate migration as an opportunity for development in both rural and urban areas of Bangladesh. Here I examine how this is facilitated through the production of uncertainty over the drivers of rural out-migration from the Southwest. Most significantly, migration that is said to be driven by climate change is also said to be inevitable (as discussed earlier), while migration driven by dynamics of development and land dispossession (through the transition from rice to shrimp or otherwise) suggests a different set of responses and solutions. Thus, claiming climate change as the driver of rural out-migration without reference to other drivers impedes possible responses to, and thus entrenches, rural dispossession.

A research institute in Dhaka that is widely celebrated by donors and practitioners and has a prolific record of publications and presentations concerning climate migration offers an instructive example of this process of knowledge production. The work of this institute to propagate discourses surrounding climate migration serves to produce uncertainty about the nature of ecological change and the demographic shifts that accompany it. Discourses about climate migrants are born out of the production of uncertainty over what constitutes "climate migration" as opposed to migration taking place for economic or other reasons. This institute regularly presents projections that between 2011 and 2050, as many as 16 to 26 million people will migrate from Bangladesh's coastal zone due to climate-related stressors. These figures are cited widely among donors, development practitioners, and policymakers in Bangladesh as the best projections for understanding climate migration from this region (which is often cited as the region that will experience the effects of climate migration most dramatically).

However, this projection, as well as the particular data analysis on which it is based, turns on the production of uncertainty about the drivers of migration, a process that, in its clear intentionality, might be appropriately called obfuscation. In one public presentation of their findings, researchers from the institute acknowledged that only 10 percent of their respondents attributed the primary reason for their migration to climatic stress. This finding is apparently at odds with their conviction that climate stress and migration are definitively linked. However, they explained that climate migrants may not be aware of the role that climate change plays in their decisions to migrate. Therefore, an alternative methodology, wherein migrants were asked what climate-related stressors they had experienced, suggested that any migrant who had experienced a "climate-related stressor" was a climate migrant, regardless of their own attribution of what had driven this decision. Given that almost all social and ecological dynamics in the coastal region of Bangladesh today can be called "climate-related," the result of this methodological decision is that almost any migratory event today can be considered a climate migration. These researchers explain that in the absence of sufficient information to differentiate between possibly climate-related drivers of migration and non–climate related drivers, it makes sense to consider all possibly climate-related migratory events to be climate migration. The result of this classification is that migration comes to be seen as inevitable as opposed to the result of ongoing local decisions driving the contingent process of agrarian dispossession. Thus, what appears as uncertainty indicating inevitability is actually obfuscation leading to dispossession, which is produced through ongoing decisions and interventions. This strategic methodological shift was confirmed in multiple conversations that I had with separate members of the research team. In chapter 5, I examine an example of out-migration from a village in Khulna that highlights the implications of this methodological choice for a particular rural community.

The methodology through which these figures were derived is based on the deployment of normative development objectives in place of a clear investigation of available data. In this way, studies of migration come to be shaped by the researchers' understanding of the fundamental desirability of migration. Another researcher affiliated with this institute asked a group of development practitioners at a private meeting in Dhaka, "If we are going to promote [migration] as a climate change adaptation strategy, how are we going to do it? Do we have enough research evidence?" In this way, establishing the "evidence" that might be used in service of these goals becomes the driver for how research and analysis is carried out.

The implications of these methodological decisions for development policy are significant. Indeed, the effects of climate change are seen as intractable and beyond the control of local development policy. Therefore, if people are migrating due to climate change, there's nothing that can be done about it. That these migration patterns will only grow in size as the effects of climate change increase is assumed to be self-evident. Moreover, the need for people to escape these climate-vulnerable communities becomes equally self-evident. This uncertainty over the exact drivers of migration is ubiquitous in research examining climate migration around the world. The inability of researchers to ascribe any particular changes at this spatial or temporal scale to climate change leads to understandable uncertainty in drawing causal relations between climate change and multicausal patterns such as migration. There is nothing fundamentally flawed methodologically about attempts to explore patterns in the face of such gaps in information. However, it is necessary to be aware of the normative claims that may result from such assumptions, as well as particular practices that produce uncertainty about these dynamics. If, for example, these migrations were understood to be the result of the expansion of shrimp aquaculture (as explored in chapter 5), then the potential responses would be much more diverse, and would perhaps seem less inevitable. The value of the landscape, and of the communities that inhabit it, thus shifts along with the uncertainty over the biophysical dynamics shaping them now and in the future. Practices of uncertainty about climate change and environmental crisis in coastal Bangladesh shape new terrains of possibility for production and accumulation in the region. Within these politics of knowledge production, uncertainty about ecological change is claimed, produced, and mobilized to pursue particular visions of developed futures.

Yet, as I have highlighted in this chapter, these categories of certainty and uncertainty are profoundly unstable. Both are subject to interpretation and manipulation and always in flux. The effects of popular discourses surrounding Bangladesh's dystopic climate future both within and outside the country are embedded in the production of scientific knowledge about the future of this region, which is seen as under threat. Thus, these politics of uncertainty both shape and are shaped by the adaptation regime, constructing limits on the region's possible futures.

4

THE SOCIAL LIFE OF CLIMATE SCIENCE

Circulations of Knowledge and Uncertainty in Development Practice

> "We appear to be describing the same episode, but within that episode we see different actors and different social relations."
> —E. P. Thompson, 1975

The politics of uncertainty are embedded not only in the production of scientific knowledge about ecological change in Bangladesh, but also in the development interventions that respond to and produce those changes. In Bangladesh, new programs are emerging in the name of development and adaptation that govern normative visions of what can and should be produced, which are rooted in particular understandings of the ecology and how it is changing. Programs that propose a transition from rice agriculture to shrimp aquaculture as a climate change adaptation strategy are predicated on a politics of knowledge through which the possibility of ecological crisis becomes an opportunity for accumulation through agrarian transition. The environmental degradation and agrarian dispossession that ensue from this transition to shrimp are linked with normative ideas about what "development" might look like in this region in the time of climate change. In this chapter, I examine these politics in the context of three key development discourses and planning processes: interventions surrounding waterlogging, the promotion of "climate migration," and land use zoning practices in the coastal region. Each of these examples highlights the instability of the categories of certainty and uncertainty within development-planning processes, and thus illuminates how the politics of uncertainty are manifested in Bangladesh today. In highlighting this shifting relationship between certainty and uncertainty, these examples also draw our attention to how and under what conditions claims about uncertainty are made and to the interests these claims ultimately support.

The diverse, intertwined drivers and dynamics of change are defined by a sense and discourse of *uncertainty* over their relative importance in relation to one another—what changes they will cause, how they will do so and under what time frames, and how they are interrelated with one another. This uncertainty is so central to the discussion of this region that it has itself become an important driver of change. While the specter of climate change and uncertainty over its impacts on the future draw increasing amounts of attention to the region, they also draw attention away from the region's history and political economy and their ongoing relationships with ecological transformation. To be clear, the uncertainty that marks the decisions about how to manage this space is real. But the way in which it is claimed, generated, and mobilized toward particular futures is marked by inequitable power relations. These unequal relations of power give shape to the politics of uncertainty.

Along with the insights of the previous chapter, in this chapter I go further than arguing that scientific knowledge about environmental change in Bangladesh is socially constructed. Rather, I demonstrate that neither nature nor social power operate independently of one another; they are always co-produced (Jasanoff 2004). How this landscape is changing cannot be understood independently from historical and present efforts to know and change it. The very science that seeks to understand these patterns both shapes and is shaped by them.

Discourses and practices of uncertainty in relation to climate change are often associated with inevitability, though they are distinct from one another. This relationship between uncertainty and inevitability in the context of climate change is paradoxical, since they would seem to be opposites. However, uncertainty about the relative weight to assign to climate change in understanding contemporary socioecological transformation is embedded with an assumption that the weight of climate change will grow as the climate crisis worsens. In the absence of greater contextual information about other drivers of change and how they might be mitigated, the growth of this crisis comes to be seen as inevitable. Only by texturing our understanding of socioecological change with a more detailed analysis of the ongoing and contingent social, economic, and political dynamics shaping these transformations can this teleology of inevitable climate crisis be challenged. In what follows I examine uncertainty as a politics and a set of practices, in both discursive and material formations. I focus on these practices to draw attention to the ways in which uncertainty takes an active, rather than passive or incidental, relationship to knowledge. Modes of governing emerge out of interactions with uncertainty by particular actors and through particular epistemes.[1]

The politics and practice of uncertainty take many forms, of which I am concerned particularly with the following: (1) How uncertainty is *claimed*: By this I refer to the interface of science and politics—why and how particular uncertainties are identified. (2) How uncertainty is *generated*: Here I refer to the active production of a sense of uncertainty through ignoring or undermining that which is already known but is seen as politically inconvenient. (3) How uncertainty is *mobilized*: This concerns the politics of what is done in the name of uncertainty and how it is used to pursue particular ends in policy and development. In what follows, I explore each of these interconnected practices of uncertainty and the ways in which they shape the dynamics of production and social reproduction throughout southwestern Bangladesh.

Understanding and Addressing Waterlogging

Though recent years have seen a significant increase in attention to the Southwest and the particular ecological challenges the region faces today, these concerns are not new, and neither are interventions seeking to address them. For the past twenty to thirty years, donors in Bangladesh have provided significant humanitarian relief funds for communities in the Southwest affected by seasonal and persistent waterlogging, assistance that has grown steadily in its scope and magnitude. While *waterlogging* essentially refers to recurring and/or long-lasting flooding, as will become clear, determining exactly what defines waterlogging—its conditions, drivers, and temporalities—is not a neutral question. One European donor estimated that in the previous few years preceding 2015, approximately 30 million euros (about $36 million) had been spent on humanitarian aid related to waterlogging, and remarked, "What do you get for that? Well, you get a few people who have been kept alive, they might have a few more tarpaulins." In 2012, a group of these donors, motivated by a growing demand for funds to address this chronic crisis and emboldened by "resilience" frameworks, which they said helped them to examine overlapping development challenges and objectives, together set out to investigate the root problems of waterlogging in the Southwest. This investigation, which was managed by the Food and Agriculture Organization of the United Nations (FAO) Bangladesh office on behalf of this donor consortium, came to be known as the "Mapping Exercise on Water Logging." Waterlogging is a specific and concrete problem, which in Bangladesh is often claimed in popular and development policy narratives to be a particular manifestation of sea level rise. Yet it also sits at the intersection of a wide range of social and ecological issues, and thus the mapping exercise took up waterlogging as a kind of proxy for investigating these compounded concerns throughout the region.

The discursive and epistemological morass that grew out of the mapping exercise is an object lesson in the politics of uncertainty in Bangladesh, particularly in relation to the practice of claiming uncertainty within the adaptation regime.

In spring 2014, when I began asking questions about the study in private conversations and interviews with donors and consultants participating in the process, it was already clear to all involved that they had inadvertently wandered into a tense and awkward political terrain for which, a draft report claimed, "no elegant solution exists" (FAO 2015, 92). "Elegance" here is a euphemism for political convenience, not necessarily for scientific or physical capacity. At the heart of this tension are the paired debates over what exactly is causing waterlogging and, subsequently, what the solution is. Much of these debates are motivated not by a lack of information as such, but by disagreements over how to interpret its implications, and thus how the problem should be presented. Nevertheless, claiming uncertainty allowed the actors involved to avoid directly addressing the interests served by these negotiations.

The mapping exercise motivated a series of uncomfortable conversations about what is really driving waterlogging and disagreements over identifying the actual problem. Though the sedimentation of rivers and subsidence of the land inside the polders has facilitated waterlogging, impediments to drainage are usually related to conflicts between rice agriculture and shrimp aquaculture. Specifically, irrigation and drainage canals are often blocked by shrimp producers in order to keep saltwater inside embankments, and the canals themselves are often seized for use as shrimp *ghers* (ponds), preventing farmers from flushing out stagnant water.[2] The land use and tenure arrangements involved in these conflicts are marked by tremendous inequity, compounded by the fact that in most cases those involved in the occupation of land and canals for shrimp cultivation are politically and economically influential and enjoy the support of political elites and often NGOs as well.

The political economy of land tenure and water management in these communities thus drives the waterlogging problems they are experiencing. It also shapes the way in which the issue is understood. While waterlogging is a significant problem for some—specifically agriculturalists, day laborers who were dependent on rice cultivation, and those displaced from homes in lower-lying floodplains—for those who benefit financially from the expansion of shrimp, it is an opportunity and a boon. These tensions over the normative value of flooding have a long history in the region, as described in chapter 1. Specifically, drainage came to be understood as a problem in the colonial period in cases where it impeded attempts to meet revenue demands (D'Souza 2015). Thus, while waterlogging has been identified as an intractable humanitarian crisis, the mapping exercise revealed that challenges to addressing these underlying political economy issues impeded the search for a solution. The conversation over waterlogging itself became contentious because it entailed identifying political issues that donors and policymakers did not want to discuss openly.

One consultant acknowledged to me that the "big issue" was how to have a "straightforward" conversation, "because this is not about rational solutions to water logging. This is about who owns what." Indeed, the mapping exercise revealed disagreement over whether waterlogging itself is even a problem; for example, one participant told me, "our aim is not solution minded," explaining that they intended to address "the negative consequences of waterlogging, not waterlogging itself, per se." I understood this to mean that in the course of their investigations, the donors had found that confronting the root causes of waterlogging would require grappling with complex political economy issues which they were not prepared to address. Consequently, they decided to retain a common account of uncertainty about the problem and instead work to manage the fallout of not addressing those political economic dynamics.

In the presence of this inconvenient information, the practice of claiming uncertainty became a primary strategy through which donors managed this uncomfortable political landscape. Despite recognition among most if not all participants that the political economy was driving the ongoing waterlogging issues, they retained a shared public account of uncertainty, which facilitated their continued efforts. Even as the research conducted for the mapping exercise illuminated the role of shrimp aquaculture in causing and exacerbating waterlogging, claims of uncertainty coupled with references to climate change allowed donors and practitioners to avoid addressing these apparent drivers of change. While donors continued to refer to their associated activities in the Southwest as "climate-related," if not specifically designating them as climate change adaptation projects, in private conversations they often questioned this official narrative, clearly recognizing nonclimate drivers of the changes they were confronting. One donor told me about a private meeting among European Union (EU) member states in Bangladesh to discuss waterlogging and related ecological challenges at which participants discussed the important role that shrimp aquaculture plays in degradation. According to this donor, the discussion involved an examination of whether the fact that the EU imports roughly 50 percent of Bangladesh's shrimp exports means that the EU, both through trade and support through development programs investing in aquaculture expansion, bears some responsibility for ongoing environmental transformations resulting from shrimp cultivation. However, this analysis cannot be found in any public reports or statements. Thus, both the conversation at this private meeting and the donor's comments to me afterward, represent moments of fissure in these practices of uncertainty.

The draft report of the mapping exercise languished as a "Draft for Consultation," with some participants saying it would be referred to as a draft indefinitely to allow space for this type of equivocation (the classification itself thus becoming a tool of uncertainty). Nevertheless, it provided a roadmap for the planning and coordination of activities among concerned agencies in subsequent phases of the mapping exercise. Toward the end of the report there is a table mapping the responsibilities of various agencies and activities that could be undertaken to fulfill them, which summarizes the key conclusions of the exercise overall. The table, which was regularly referred to by participants as "the matrix," reflects these fundamentally contradictory activities and objectives. This is manifested most clearly in its simultaneous recognition of aquaculture's role in waterlogging along with proposals for facilitating aquaculture expansion. For example, in recognition of the government's role in facilitating waterlogging in shrimp cultivation, one column of the matrix indicates the need to curtail waterlogging by removing illegal dikes in canals and small rivers that are used to create enclosures for shrimp. However, elsewhere in the matrix we see policies proposed to do precisely the opposite through the development of new aquaculture technologies and other types of support from NGOs or UN agencies for the expansion of shrimp cultivation. These contradictions reflect disagreement over what the problem is and tensions over mandates and conflicting goals. Instead of confronting these contradictions directly, claiming uncertainty about the drivers of and responses to waterlogging serves to obscure them. Claiming uncertainty in this context, then, serves to entrench existing politics of resource distribution in rural communities, even while these policies are claimed to be fundamentally apolitical.

These contradictions are inherent in the approach of the report itself, which explicitly avoids the politics driving the waterlogging crisis in the Southwest. "The matrix does not analyze," one consultant said to me, meaning that comprehensive analysis to reach a deeper understanding of the drivers of this waterlogging was not an expected or intended outcome of the exercise. Ultimately, the report does not provide a coherent vision for eliminating the waterlogging problems that the mapping exercise theoretically aimed to address in the first place. By retaining a collective account of uncertainty concerning the drivers of waterlogging, the project thus avoids taking steps toward mitigating them.

The mapping exercise illuminates the role uncertainty can play in the process of what Tania Li calls "rendering technical," referring to the enframing of a field of intervention appropriate to the kinds of solutions that experts have to offer (2007). The mapping exercise rendered technical the problem of waterlogging by defining the problem only in relation to available solutions. Uncertainty was employed to evade available solutions that were deemed to be undesirable. As the problem of climate change is rendered technical, the constant production of scientific knowledge about its effects in Southwest Bangladesh produces a particular field of intervention. Thus, where the adaptation regime limits the scope of possible strategies for addressing these changes, uncertainty is deployed to delimit a vision of possible futures for the region.

I gained firsthand insight into how these competing concerns involving waterlogging are negotiated through a project managed by the United Nations Development Programme (UNDP). I had originally gone to meet with a staff member at the UNDP Headquarters in Dhaka who was involved in the mapping exercise, to learn about the UNDP's work with the study and interventions designed to mitigate the impacts of waterlogging. While I was there I learned about a pilot program they were calling "TRM++," which they planned to implement in the Tala subdistrict of Khulna Division (Tala is bordered to the east by the subdistricts of Paikgachha and Dumuria, locations of field sites explored in greater detail in chapters 5 and 6). TRM stands for Tidal River Management. It refers to a vaguely defined set of strategies for mitigating waterlogging impacts within the polders, which are broadly targeted at temporarily opening up sections of embankment to selectively reintroduce tidal flow. The goal is to raise the land elevation by facilitating the accretion of sediments on lowlands through intentionally flooding them with sediment-rich tidal water.

The history of TRM strategies is generally traced back to an incident in Beel Dakatia, a marshy lowland in Polder 25 that in 1982 began to suffer from drainage congestion and chronic waterlogging related to the systemic engineering problems with the polder system described in chapter 1 (Rahman 1995).[3] By 1990, over 16,000 hectares of land and homesteads in Beel Dakatia were continuously under water (Adnan 2009, 113). In response, local farmers, having realized that the embankments were preventing tidal sediments from being deposited on and building up their land, organized a mass mobilization to breach the embankment and allow the water from the river to flow inside the polder. Their strategy was at least partially successful: with tidal activity resumed, sediments flowed freely into the polder, rapidly building up the land. By 1992, at least 2,500 acres of land had resurfaced, and farmers had resumed rice production (Adnan et al. 1992, 52). Shortly thereafter, however, the government stepped in with a new intervention to address the infrastructure and drainage issues in Polder 25, which involved plugging the cuts that farmers had made in the embankment.

Subsequently, TRM has gained a kind of cult status, being celebrated as a potential remedy for problems created by the polders that relies on unique local knowledge (Hossain, Khan, and Shum 2015; Khadim et al. 2013; Shampa 2012). The pioneering work of farmers in Beel Dakatia to forge this experiment makes it a popular idea with activists, while it is also occasionally discussed in development circles as a potential compromise to address the problems with the polders without completely dismantling them. However, this celebration of the potential of TRM notwithstanding, its widespread adoption in development planning has been thwarted due to challenges of replicability. Not every waterlogged area is physically suitable for TRM, which is dependent largely on the proximity and path of the river channels in relation to the area to be raised (if the water has to come too far or take a circuitous route to the lowland, the sediment will not make it all the way, and therefore the land will not be successfully raised). The process of translating the TRM experience into transferable engineering principles has also jettisoned the social and political particularities that made the experience in Beel Dakatia successful (Shahidul Islam and Kibria 2006).[4] Its implementation is messy and very slow, requiring concerted efforts at collective deliberation and action. These characteristics make TRM difficult to translate (particularly by community outsiders) into concrete development programs with strict and limited time frames and predetermined goals and objectives. Specifically, farmers in Beel Dakatia had forged a consensus on returning their land to rice cultivation, and TRM facilitated the change in water regime that was required. It was this work of social mobilization that made their efforts successful more than any particular innovation in water engineering. However, in other contexts, a lack of consensus on what to produce may also involve a lack of consensus on the amount and scheduling of water availability. As noted in the discussion of the mapping exercise, those who were interested in cultivating shrimp may regard as favorable the same flooded conditions that rice farmers regard as inimical to production. Moreover, attempts of development agencies to put TRM into practice have been confounded by the challenges of determining whom to compensate for lost production revenues during the period during which the land is forcibly inundated, and how (van Staveren, Warner, and Khan 2017). These challenges of pursuing TRM in particular places have discouraged researchers and development agencies from pursuing TRM rather than water management strategies with broader potential for replication by external actors (Auerbach et al. 2015b).

The UNDP's TRM++ project was intended to address these deficiencies in previous attempts at replication. The program's official documentation listed the intended outcome of the program as follows: "By 2016, populations vulnerable to climate change and natural disaster have become more resilient to adapt to risks" (UNDP 2014, 1). In the early stages of implementing the project, UNDP staff in Dhaka asked me to go visit the village where it was being implemented, as they were looking for assistance in documentation and analysis of the process. Extolling the project's innovative features before I went to see for myself, the UNDP staff told me at length about the "community consultation process," which they said was focused on the landless and the poorest villagers. They described their intervention as primarily "social mobilization" with "some physical aspects." The idea was essentially to choose a suitable TRM site, and then through consultation with the community, to financially compensate the landowners for the crop loss during the seasons when the land had been intentionally flooded, but also to allow the poorest community members (landless people and sharecroppers) to fish in the flooded wetland while it was under water. Thus, the landowners would be compensated for the lost earnings potential from their crops or land rents, while the agricultural laborers would be compensated for their lost wages through earnings from fishing income. One staff member explained to me, "Our key interest is around the social engineering of it," and in demonstrating to the government that they could "get more social feasibility out of TRM." While I was apprehensive about the conflation of "social engineering" and "social mobilization," as I understood it, the goals of the program were to address the environmental impacts of the polder system with sensitivity to their intervention's social impacts. I agreed to go see for myself.

UNDP had commissioned physical, environmental, and socioeconomic baseline studies of the proposed project sites prior to initiating work there, with the studies to be carried out by faculty from Dhaka and Khulna Universities (Khulna University 2014; University of Dhaka 2014). Having read their richly detailed reports before arriving, I had learned that there were two sites being considered for the TRM++ project. The physical baseline study clearly demonstrated that one of the proposed sites would be a suitable option for implementing TRM, as it was located directly adjacent to the river, and the velocity of the water flowing in from the river would carry a substantial amount of sediment to the currently waterlogged lands. These are the ideal physical conditions for successful implementation of TRM. The other site under consideration, according to the report, was not suitable for TRM, as its physical geography separated it from much of the tidal influence. It was located approximately 840 yards from the river, and water would be forced to travel to the site through a narrow canal with several angles, all of which would reduce the velocity of the water and its capacity to carry sediment to the beel (University of Dhaka 2014, 26). However, the reports also indicated that while local agriculturalists (landless laborers, sharecroppers, and landowners) at the suitable site were in favor of TRM as it would restore their lands for rice cultivation, the plan faced opposition from a small group of elites, who had leased the waterlogged lands (in some cases forcibly) to cultivate shrimp. These people, who are referred to in the reports as gher businessmen or gher owners, opposed the project because raising the land to make it viable for agriculture would undermine their ability to continue cultivating shrimp (Khulna University 2014, 38). As observed in the mapping exercise, for some people in this community, waterlogging was a threat to their livelihoods, while for others, it was a source of economic opportunity. The site that was unsuitable for TRM did not suffer from these same conflicts because there were no ghers there. Being so far from the river, it was difficult to transport sufficient saltwater there for shrimp cultivation. This made agriculture the only option, irrespective of land tenure arrangements.

Given this clear assessment of the technical feasibility of the intervention, when I arrived in Tala I was surprised to find out that only the site determined unsuitable for TRM had been selected for its implementation. The local staff explained to me that they had decided not to pursue the work in the suitable site because "there was a lot of complexity." As I probed further, I learned that there were about a hundred landowners in the suitable site, all of whom were in support of the TRM intervention, but there were also about a dozen gher businessmen who were not landowners but who had leased all the waterlogged land for shrimp cultivation. "The businessmen didn't want to stop their gher business. They were also very influential," one of the project staff told me. Faced with this opposition from influential businessmen, the project staff decided to give up on the suitable site and to redirect their efforts to the other site, despite the prior determination that its physical conditions made it unlikely that TRM would succeed there. "Ok, it is true," the staff member conceded to me, "the canal will get silted up [again] within a few years, so it is not a permanent solution." Nevertheless, the TRM intervention would be more politically convenient there because there were no shrimp businessmen opposing it.

Before I went to visit the project site, the staff told me about Bablu, a resident of the village who was described in turns as a local farmer and the founder of a social welfare NGO that sought to help the village's poor residents. Because of his position, he was the primary contact for the project staff and was organizing local residents who would become its beneficiaries. When I met Bablu on the way to the beel, I immediately noticed his shiny watch, his unblemished attire (including pants rather than the traditional *lungi* sarong of a farmer), and his motorcycle. Through these trappings, along with his confident demeanor, there could be no doubt that he was not a member of the landless population I had been told about before I came, who were participants in the project's "social mobilization" efforts. Both Bablu and the program staff were clearly unsettled by my request to spend the day alone, walking around the village and talking to people. They wanted to serve as intermediaries and did not think more than half an hour would be necessary for me to see everything. It shortly became apparent to me why this was the case, as my conversations in the village revealed serious discrepancies between the project's implementation and its design as it had earlier been described to me. The inconsistencies were clearest when I crossed the canal to a patch of land where I found the humblest homes in the village. While the beel itself was flanked by *pucca* (brick) houses on large plots, often with gardens, the houses on the other side of the canal were all constructed of mud and corrugated metal and were in various states of disrepair. This was where the landless people lived. Most of them worked as agricultural laborers and supplemented their income with other migratory work, such as working in brick fields and collecting honey and other resources from the Sundarbans. A clear consensus emerged as I walked from house to house talking with these landless residents—most of them had not heard of TRM or of the UNDP project. "No, they haven't talked to us," one man explained to me, adding, "the NGO people come talk to the *boro lok* [elites], then they make decisions, do whatever they want." Those who had already heard about the TRM project and those whom I told about it were convinced that the poor would not be its beneficiaries, as had been described so emphatically at the UNDP office in Dhaka. When I asked about Bablu's social welfare work, several of these residents laughed at me, and one man explained, "He doesn't help the poor; he is just a large landowner."

This exclusion of the village's poorest residents was confirmed at the end of the day when the UNDP staff arranged a meeting for me with the people they described as program beneficiaries taking part in the social mobilization process. Of the approximately twenty-five beneficiaries present, none of the landless people or day laborers I had met during my day in the village were in attendance. I asked those present to go around the room and tell me what kind of work they did, and it became clear that they were all landowners, among the most elite in the village. Several of them were businessmen, including two jewelry store owners and the proprietors of a butcher shop and a convenience store. As we discussed the plans for economic activity in the beel during the wet season, when it would be flooded through the TRM intervention, they said they planned to lease it out for others to fish in. When I asked what they thought the village's landless people would do during this time, they looked back at me blankly. They could take microcredit loans, one man suggested.

It would be easy to dismiss this as a case of faulty program implementation and the elite capture of a development intervention. However, the decision to reject the first site, which was suitable for TRM, and the role of Bablu and wealthy landowners in the second site suggest that accepting (and even entrenching) the power of local elites was foundational to the program itself. Despite obvious concern in the program's conception and formation for the impacts of waterlogging on the poorest, program staff repeatedly declined opportunities to directly address local power imbalances. In refusing to unsettle and ultimately choosing to work within these existing power structures, the program failed to mitigate the effects of these unequal agrarian political economies, despite the obvious interest of the poorest residents of both communities.

A couple months after my visit to Tala, I saw a Bangladeshi acquaintance who works at a university in Dhaka and has also conducted research in Paikgachha. He told me that he had been hired for a consultancy to evaluate the TRM++ project. I told him that I had visited the project site in Tala myself and began to convey some of my concerns to him. He interrupted me, saying he did not want to know what was wrong with the project. Reminding me that many faculty members at Bangladeshi institutions are paid so little that they depend on consulting contracts to supplement their livelihoods, he told me that my perspective on the problems with this project was a luxury he could not afford. He said he felt that if he articulated these criticisms about the program, he might not be hired for further consulting contracts in the future. When I pressed him on this, saying I thought we both had a responsibility to share with each other and to ensure the transparent dissemination of information about development interventions in our shared field sites, he abruptly got up and left the room.

In the years following this interaction, I have had the opportunity to parse it with several colleagues, including other scholars and activists both within and outside Bangladesh. Their responses have varied widely from sympathetic to highly critical of my fellow researcher, reflecting the complexity and ambiguity surrounding the role of individual actors in shaping these politics of uncertainty. The adaptation regime is a system of governance composed of individual actors making decisions that are at once collective and individual. They exercise agency within the constraints of multiscalar power structures within which they possess varying degrees of power themselves. What are the roles and responsibilities of these individual actors in negotiating or contesting these politics of uncertainty?

On the one hand, this incident speaks to the unequal structure of research funding and compensation, which is fundamental to the dynamics of knowledge production in Bangladesh today. Whether one has the ability to say certain things and even know certain things about socioenvironmental change is profoundly shaped by one's position in relation to these structural power dynamics. Yet the exchange also speaks to the ways in which this political economy of knowledge production fundamentally shapes landscapes, often promoting the interests of elites at the expense of the poorest. In this case, the active refusal of knowledge (on the part of the UNDP staff as well as the consultant tasked with evaluating the project) about how this intervention may serve the interests of elite shrimp gher operators supported those interests at the expense of the farmers and laborers for whom waterlogging is the source of much suffering. Assessing these different analyses of this incident in its aftermath forces us to grapple with the messy everyday politics of the circulation of knowledge and uncertainty about environmental change and adaptation in Bangladesh today. The roles and responsibilities for shaping and contesting these politics of the adaptation regime are often ambiguous, with actors exercising power individually though never autonomously.

This example of the TRM++ project speaks to a wider concern with the politics of adaptation and with the exercise of knowledge about designing and implementing effective intervention. As the diverse experiences in Beel Dakatia and Tala demonstrate, TRM itself, as with any technical intervention or adaptation strategy, does not possess a fundamental politics. TRM is shaped by the communities where it is practiced and the relationships within them, by whose interests it does and does not serve. It is only through decisions about how to respond to these dynamics that the politics of TRM are manifested. In choosing to ignore the role of these socioeconomic factors, the program also failed to effectively address their environmental consequences. As the knowledge about the potential for TRM in Tala traveled (from the historical experience of Beel Dakatia through the baseline studies carried out by researchers at Dhaka and Khulna Universities, and then to program design and implementation), details about what success would look like and how to achieve it were often eliminated or transformed. Through this process of translation, uncertainty was mobilized to shape the use of TRM in a particular way, ultimately failing to address the social and economic factors that might have made the TRM++ approach successful in mitigating both waterlogging and inequality together. These same politics are negotiated in every adaptation intervention, either contesting or reproducing these power structures in different ways.

Rice and Shrimp Conflicts

As uncertainty about ecological change in the Southwest is often claimed, it is also actively produced through particular development discourses and research practices. Though claims to uncertainty as described earlier also clearly serve a purpose, in this section I refer to those practices through which uncertainty is generated through the act of knowledge production itself. Uncertainty about conflicts over shrimp production and uncertainty about rural out-migration are two areas where these practices are manifested most clearly, and I turn my attention to them next.

While donors, development practitioners, and researchers claim uncertainty over the interface between the ecological impacts of shrimp aquaculture and those of climate change, when it comes to the social implications of shrimp aquaculture, uncertainty is actively produced by many of these same actors. Often this production takes place in the interstices of claims to authority over decision making and gaps and evasions of translation between local and international actors.

Though the conflicts between shrimp and rice are usually ignored by concerned donors and policymakers, when they are raised, these actors respectively demur, claiming a lack of responsibility or influence over the political economy of production in the Southwest. Donors and foreign development practitioners claim that knowledge and responsibility concerning these conflicts lies with the government, while civil servants working for state agencies claim it is for the donors supporting aquaculture programs to investigate and address such claims.[5] In one interview with an official at the Bangladesh Department of Agricultural Extension (DoAE), I was told that the DoAE does not get involved in conflicts over rice and shrimp unless farmers are interested in making the shift to shrimp, in which case it may provide support to do so. Otherwise, it is the responsibility of donors to assess the costs and benefits of shrimp production, as "assessment" requires money, which donors manage and allocate. The official claimed to me that the DoAE possesses technical expertise in production technologies but does not have additional expertise in the kinds of "assessment" that require economists, sociologists, and other experts, work that he believed donors are uniquely suited to support and initiate. The official cited the mapping exercise as an example of donors funding and making decisions about what will be examined in regard to ecological change and development, as well as how such assessments will be carried out. In this sense, the official highlighted how decisions over what will be understood and what will remain uncertain are shaped in the development-planning process.

For their part, donors uniformly dismiss responsibility for evaluating or addressing the conflict between shrimp and rice. In one interview, a foreign development practitioner involved in a major USAID-funded shrimp aquaculture program referred to these pervasive conflicts as "just personal local fights." In another conversation, when I asked a question about "land grabbing" for shrimp production, another practitioner corrected my phrasing by saying "you mean 'multiownership.'" I subsequently heard this linguistic conceit of "multiownership" used repeatedly—it is seemingly aimed at obfuscating the power dynamics embedded in these "multiple" claims to land at the frontier between shrimp and rice. This diminution of the significance of these resource conflicts and the role they play in the expansion of aquaculture production reflects the selective mobilization of knowledge about land relations that is embedded in programs to support shrimp cultivation.

Development practitioners recognize these conflicts and their significance with varying degrees of concern, though with a common refusal of responsibility. One senior diplomat from the United States acknowledged the land conflicts related to shrimp while also seeming to absolve his own office's support of the aquaculture expansion, exclaiming, "Well, the mafia down there has just messed everything up!" By "mafia" he referred to the reports of land grabbing, a practice he intended to claim was not inherent to shrimp production, which could be, he said, a billion-dollar industry if not for the malpractice of this isolated cabal. By refusing to acknowledge the complicity of aid agencies (indeed those supported by the United States government itself) in rural dispossession through shrimp aquaculture expansion, such comments produce uncertainty about the nature of agrarian change and its drivers.

Although the dynamics of land under shrimp cultivation in Bangladesh have long been brought to the fore by activists in public discussions both locally and internationally, development policymakers frequently refer to their lack of responsibility, certainty, or understanding of the dynamics of shrimp production to defend their support of it. The *National Aquaculture Development Strategy and Action Plan of Bangladesh*, a policy document developed by the FAO for the Bangladeshi government, provides an example of this slippage. Buried in an appendix deep at the back of the document is a "logical framework" or "logframe"[6] in the form of a table identifying the assumptions and risks of aquaculture development in Bangladesh. The table indicates the assumption that benefits of shrimp aquaculture to the poor and landless are predicated on an "equitable land and water allocation system" (37). Similarly, the risk embedded in spatial planning or mapping of zones for aquaculture development (a process examined in further detail later in this chapter) is "that potential resource use conflicts are not resolved among the various agencies" (39). In both cases, these are fatal assumptions. As one donor pointed out to me in a conversation about this plan, in a functional logframe, both these assumptions would be "deal breakers." The failure to acknowledge these flaws—specifically, that land and water allocation is not equitable, and that resource use conflicts have not been resolved—suggests a particular relationship to how information about such equity in resource distribution is assessed and incorporated into development planning. Who is responsible for addressing these concerns—and these fatal assumptions—is also, then, a matter of uncertainty. Even as the FAO itself created this national planning document, it produced uncertainty about the fundamental role of this planning in shaping the equity of land and resource distribution.

Zoning

One of the key tools cited by development practitioners in Bangladesh as essential to climate change adaptation is land use zoning in the coastal region. Referred to by donors, NGOs, and various government line ministries in planning documents, project proposals, and in regular conversation among staff at these agencies, zoning is celebrated as a kind of panacea for all climate-related planning challenges. It is said to resolve uncertainty by rationalizing an otherwise unwieldy planning process, concretizing planning through the use of hard-nosed scientific data to objectively reconcile disparate aspirational visions of development.[7] Zoning is said to maximize economic benefits and to reduce conflicts of all sorts—between forests and human settlements, rural production and urbanization, and, critically, between rice and shrimp. In this sense, zoning serves to exculpate the promotion of shrimp aquaculture from critique of its negative impacts, and to insulate it from resistance.

Yet for all the celebration of zoning as the solution to a variety of adaptation planning dilemmas, further details about the process of rationalizing what should be grown where (and why) proved to be unexpectedly hard to come by. Over the two years during which I conducted concentrated fieldwork in Bangladesh on this topic, detailed accounts of the process of land use zoning became the veritable white whale of my research, a supposedly objective but ultimately furtive science with elusive and enigmatic methodologies. My adventures chasing the secrets of this science, I believe, ultimately taught me more about the nature of the development-planning process than any particular zoning map or methodology could have.

Indeed, despite all the times zoning is invoked to explain away criticisms of conflict, the practitioners and policymakers I talked with only described its use in this ideological sense. Zoning measures are not implemented as strict land use policies, nor are existing plans detailed enough to be put to this further use. Several different government agencies and NGOs have their own unique zoning maps, which are held internally but not necessarily shared with other agencies. An official at the Water Resources Planning Organization (WARPO, a government agency tasked with high-level water management planning) told me that he knew of distinct zoning maps produced by the NGO WorldFish, the FAO, the Bangladesh Ministry of Land, and an independent group of university researchers. Yet, while each of these agencies may have used these maps internally to justify where they would carry out their own projects,[8] they weren't agreed upon between groups or among local stakeholders and they often weren't shared publicly.

The official at WARPO described these zoning plans as necessary for curbing the incursion of shrimp into lands still suitable for agricultural production, but also potentially futile. The national Coastal Zone Policy indicates zoning as the means of resolving conflicts between rice and shrimp. Yet "it is almost too late," he told to me, inferring it may not be possible to control the conversion of land suitable for rice agriculture to shrimp aquaculture through such governance measures. He continued, "Everybody is afraid. Even the Land Ministry is skipping the responsibility. They are afraid of the moneyed people. Or they are benefitting from that money themselves." In this way, he explained, the only role of the zoning documents would be to implicitly validate the expansion of the shrimp cultivation area, in the sense that the zoning maps will not restrain aquaculture, but they may condone it.

This is reflected in the Land Ministry's own zoning map for the Paikgachha subdistrict, which the WARPO official shared with me privately. It indicates that Polder 23 (which has already been almost fully converted to shrimp) is zoned exclusively for shrimp, while Polder 22 (which is still entirely a rice farming area) is zoned for both agriculture and shrimp. The map thus indicates that agriculture is unviable in the shrimp zone, while suggesting that shrimp production could also be possible in the rice zone. When I pushed the official for further details on the methodology for determining land use suitability, he described with cynicism his familiarity with the process at the Land Ministry. "They were not at all serious about that," he said. "There were some fishy things going on. [They were under] very much pressure because shrimp export is so lucrative." The concerns cited by this official suggest that, contrary to being a tool of rationality and transparency, zoning is useful precisely in the opacity and uncertainty governing its production and utilization.

Though zoning came up repeatedly in my interviews with donors, policymakers, and practitioners, my inquiries into how zoning is carried out were frequently met with claims of uncertainty about who is responsible, how it is implemented, and the kinds of metrics and data involved. One informant would direct me to another, who would direct me to another, and sometimes back to the original person or, more often, a dead end. One project to establish a comprehensive adaptation plan for Bangladesh involved developing a policy for land use zoning about which no project team members I talked to could answer questions, until someone finally told me that the consultant responsible for this zoning was Dutch and based in the Netherlands, and would not visit Bangladesh for the entire duration of the project.

The development of the Master Plan for Agricultural Development of the Southern Region of Bangladesh is an instructive example. This document, which was officially produced by the Ministry of Agriculture but prepared by consultants hired by the FAO, is one of many Master Plans that are constantly proliferating among NGOs, donors, and government agencies to facilitate coordination and planning for development intervention. I met several times with a senior official involved in the development of this plan, who many had described to me as an expert in zoning. Each of the three times I went to his office, I asked him to describe the zoning process, and he repeatedly told me to come back on another occasion before he was willing to say anything about how zoning decisions for this plan were made. He ultimately told me that the four major parameters used in shaping zoning decisions include biophysical feasibility, types of infrastructure already available, participant interest, and "the market," though he noted that the market is usually the "first" priority. This account aligns with the discussion of zoning in the report itself, which says,

> For Khulna region, zoning for shrimp culture is important. This should be based on suitability conditioned by bio-physical characteristics of the land, current practices, emerging trends, long term market behavior, etc. Farmers cannot be forced to follow what to do or what not to do. But their activities in particular areas can be regulated and streamlined through measures of incentive and disincentive. (Ministry of Agriculture, Bangladesh, and FAO 2013, 68)

This note on zoning as opportunity for incentivizing and disincentivizing certain production practices is particularly important. It means that zoning can help identify areas for interventions to support shrimp cultivation (including, for example, provision of credit, subsidizing inputs such as feed and fertilizer, and facilitating access to markets and government land), while maintaining the idea that the choice of what to produce rests with individual cultivators. As the official explained to me, zoning became an opportunity not for making strictly scientific decisions about what was possible to produce where, but instead for identifying where development agencies might implement programs to promote shrimp based on this series of normative judgements about the desirability of shrimp production, which were grounded largely in market logics. In this way, the zoning maps attempted to instill a sense of technical rationality in a fundamentally subjective political process.

These normative dimensions of the zoning process and the utility of zoning to development intervention were reaffirmed in an interview I conducted with a high-level government official in Khulna. After villagers in Polder 29 held a major protest against attempted land grabbing for shrimp and submitted a petition to this official asking him to intervene for their protection, I had gone to his office to ask him about the conflict (and whether or how he planned to respond). In the course of our conversation, I asked him whether the land had been zoned for shrimp or rice and if this would have an impact on his decision. In response, he said to me, "In your country, you have a lot of rules and laws. In our country, we have traditional ways of doing things. Like traffic laws—we do as we have always done. We use [zoning] publicly, but in process, we don't actually use it. Zoning is for planning. The government's role is to support the private sector with exports." As this official's statement suggests, far from being a neutral science grounded in the analysis of diverse and changing ecologies, my conversations with these practitioners and policymakers revealed the normative dimensions of zoning, which involved ideas about ideal production practices and related development trajectories for the region.

Indeed, these conversations about zoning indicate that often the opinions of development and planning officials that shrimp *should* be produced were conflated with the idea that shrimp was the only thing that *could* be produced (therefore justifying zoning particular areas for shrimp). Likewise, that which *is* produced, where shrimp is concerned, is usually conflated in these plans with that which *can* be produced, though the same is not true for rice, meaning that areas that currently produce shrimp are always zoned for shrimp, while areas that currently produce rice are sometimes zoned for shrimp, if zoning officials determine that shrimp production is possible (as is the case with the Land Ministry's zoning map of Paikgachha, described previously). One official told me that in some zoning efforts, areas where land is known to have been forcibly grabbed for shrimp production from rice farming have nevertheless been zoned for shrimp because "it has to be expanded because of the economy." In zoning maps produced by the CGIAR research consortium, large areas of Khulna where farmers are currently successfully producing rice have been labeled only "marginally suitable" for rice production but "suitable" or "most suitable" for shrimp cultivation.

Moreover, the interest of particular rural communities in continuing to farm rice is apparently not a determinant of the "suitability" for rice or shrimp production in their area. What, then, is the purpose of these zoning plans? Despite claims of widespread uncertainty about how zoning is determined, I found that just the *idea* that zoning exists and can serve as the basis for rational decisions conferred a great deal of power to development initiatives that claimed it. This idea performs powerful ideological work even in the absence of information about production zones and processes for determining them. Zoning is held up almost universally as a rationale for how and where shrimp aquaculture is promoted. When I asked one senior USAID official about concerns about shrimp aquaculture increasing soil salinity (thus making land unusable for agriculture), he told me that they reserve shrimp aquaculture for areas where "it is already too late," citing zoning as the process through which they determine where it is and is not "too late" for agriculture.

The disciplining work of zoning is demonstrated by a story that one former official at the Department of Agricultural Extension (DoAE) told me about a disagreement he had had with an FAO consultant from Rome. In planning a DoAE project that was being funded by the FAO, this consultant insisted that the DoAE must promote shrimp aquaculture expansion in certain parts of Khulna where people were still producing rice. The DoAE official objected to this decision because he said that farmers in this area wanted to continue farming rice. He told me, however, that the FAO consultant said that the decision to zone the area for shrimp had already been made, so the issue was not open for discussion. In this way, the zoning plans dictated a shift from rice to shrimp, despite the present or desired production practices among the area's inhabitants. Thus, the

DoAE official indicated that not only were the zoning maps generated through normative ideas about the desirability of shrimp aquaculture, they were also mobilized within the planning process to discipline the governance of future production based on these visions.

As these examples indicate, ultimately, the power of zoning is both epistemological and material. Zoning is a practical manifestation of the social relations of power promoting shrimp cultivation, while it also neutralizes potential dissent through a sense that it employs rationality to govern gaps in knowledge. The practice of zoning inhabits a space of uncertainty over the possibility of future degradation. Within that space, it governs a transformation in production relations that is seen to be a frontier of accumulation. It mobilizes uncertainties about the social and ecological future in order to frame understandings about the necessity and inevitability of those transformations. As zoning maps shape the transformation of production practices in the image of this dystopic future, they create the very conditions that they predict. In this sense, discourses of uncertainty paradoxically produce the certainty of future dispossession.

As the examples of development planning concerned with waterlogging and land use zoning highlight, the politics of uncertainty about ecological change have become a key resource in shaping new terrains of possibility for production and accumulation in coastal Bangladesh. The production and circulation of uncertainty about ecological change in the region facilitates particular normative claims about possible or desirable futures (while also serving to diminish others). Within the adaptation regime, this uncertainty is mobilized to support the promotion of shrimp aquaculture, facilitated by the instability of knowledge about how and why the landscape is changing.

In his essay "Notes on the Difficulty of Studying the State," Philip Abrams wrote, "Any attempt to examine politically institutionalized power at close quarters is, in short, liable to bring to light the fact that an integral element of such power is the quite straightforward ability to withhold information, deny observation and dictate the terms of knowledge" (1988, 62). Abrams highlights that power operates not only by producing knowledge, but also by withholding it. Like the state, the adaptation regime generates legitimacy from the disruption and omission of knowledge about the social and political structures from which it derives its authority. At the nexus of science and development policy, the governance of normative perceptions of what *can* and *should* be produced, rooted in particular understandings of the ecology and how it is changing, are shaping the landscape in southwestern Bangladesh. This dynamic is critical to the expansion of shrimp production in Khulna. The existence of social movements challenging these normative claims surrounding the landscape's production potential calls into question these politics of uncertainty.

5

AUTOPSY OF A VILLAGE
Agrarian Change after the Shrimp Boom

I came to know Arjav when he followed me out of a tea stall. We had been sitting with half a dozen others, crowded together on a few wooden benches, drinking the smoky, wincingly bitter tea with which I began most of my mornings in this village, Kolanihat. Besides the proprietor of the tea stall, my companions were men who cultivate shrimp in small ghers (shrimp ponds), mostly a few acres in size. Arjav and the others were collectively lamenting the recent fall in the price of *bagda* (giant tiger) shrimp, which was already untenably low. Between December 2014 and the following April, the price they were earning at the market had dropped from 350 *taka* per kg to 300 *taka* per kg (approximately US$4.15 to $3.50). Just to break even, cultivators need to earn approximately 500 taka ($5.90) per kg. "*Loss hoye jabe*" ("There will be a loss this year") they told me. The mood was low and indignant. The men spoke sardonically about NGO field workers who had come to provide "demonstrations" to them of proper techniques for cultivating shrimp and improving their yields. One recent method, they quipped, involved scattering tea leaves on the surface of the gher, in an attempt to kill off any lurking (and to the fieldworkers, unwanted) native species. The advice provided by these NGOs was not "practical," they explained. These are techniques, they said, that are clearly dreamed up by people who have never cultivated shrimp before.

Like the others, Arjav was frustrated about the shrimp export market, especially the losses that the gher operators would suffer in its decline. But he wanted to talk about his larger concerns about shrimp production and what it was doing to the village and its residents. In this village, rice farming has given way almost entirely to shrimp cultivation, and the once-verdant landscape is now blanketed by this new terrain of shrimp production. Arjav and I strolled slowly along a narrow path flanked on both sides by murky ghers. The late April heat (it was the hottest time of the year) was unabated by shade, as most of the village's trees had died in the aftermath of the shrimp boom. This intemperate heat frequently made long days working in Kolanihat feel oppressive, relative to the other villages in which I conducted research—both for myself (a relatively fatigable ethnographer prone to improbably high volumes of perspiration, as residents were wont to point out), as well as for laborers doing much more demanding work than wandering around chatting with people and sipping tea. These laborers often complained (far more defensibly than I) about the death of the trees, which they said used to provide shade where they could sit and take breaks from the midday heat. Without them, one man asked, "How will we breathe?"

Arjav has been participating in the shrimp business for about thirty years, he estimated. He lives in one of Kolanihat's few *pucca* (cement) houses, a relatively large one, which he told me he built shortly after the shrimp boom began in the 1980s with profits from a successful business trading shrimp fry (postlarvae). Unlike many others in the village, Arjav has benefited materially from the transition to shrimp. His sons have continued in the shrimp trade, moving to the nearby town of Paikgachha and gaining more commercial success than Arjav himself. They would not be suited to working in the village anymore, he said—they ride around on motorcycles now and are *onek porishkar lok* ("very clean people"), a phrase denoting their wealth and lack of inclination toward manual labor. More importantly, he explained, with the transition from rice farming to shrimp aquaculture, there were no more opportunities for them to work in the village so they could not have stayed. If his sons had not left, *bhat hobe na* (literally "there wouldn't be rice"), meaning they would not have been able to provide enough food for the family to eat, despite Arjav's financial success in the shrimp trade. He also explained, with a sardonic smile, that his sons were better suited to the shrimp-trading business than he was because *tader puji dorkar*, literally meaning that capital is required for this work, although also implying the necessity of a capitalist sensibility. "*Amar puji choto*," he said, explaining that he personally did not have the sensibility needed to thrive in the shrimp business. Although Arjav himself managed a relatively successful shrimp gher, the family depended on his sons' off-farm employment for daily sustenance.

Despite what would appear to be a story of success with shrimp production, Arjav tells me in no uncertain terms that he wishes they—the village as a whole—could all go back to growing rice. Sitting on the veranda of their house, staring at a stand of large bamboo rice silos in their courtyard, which Arjav and his family said had been empty for at least a decade (see figure 5.1), Arjav's brother referred

FIGURE 5.1. Arjav's empty rice silo.

Photo by the author.

to the time when rice was growing in their village, saying, "During that time, there was peace here, but now there is no peace. The peace has died." Here he echoed a broader sense of the experience of loss of the agrarian political economy of Kolanihat. While the incursion of shrimp has been associated with a very real loss of peace through physical violence perpetrated by shrimp businessmen and their hired sentries, Arjav's brother was also referring to the peace associated with other amenities of the traditional agrarian livelihood of the village—trees, subsistence production of abundant rice, fruit, and vegetables, and the agricultural landscape they once enjoyed. While it is important not to romanticize the harmoniousness of rural life in this region prior to the shrimp boom, this profound sense of loss is shared widely among the relatively privileged people like Arjav's family and the poorest families alike.

Arjav is concerned not just about the shift this has caused for his own family, but for the entire community. He described a great exodus from the village of people leaving to find work, mostly to India: "Daily, in the morning, at night, people are leaving—all the working people." This account of migration is confirmed by the rest of my research in this village and the surrounding area, as well as interviews directly with some of these migrants in Kolkata who have left this and surrounding villages. But Arjav's reflections on what this might mean for the future of the village were particularly ominous. When I asked him what he thinks the village will be like in twenty or fifty years, he told me, "If there isn't rice, the village people won't be able to stay."

Arjav's story illustrates the dramatic agrarian transitions that are taking place throughout the shrimp-cultivating region in Khulna. It highlights the effects of the transition from rice to shrimp not only for landless people and agricultural laborers, but also for small landowners who have actively participated in the shrimp boom. It also exposes the paradox that even as some smallholders have increasingly begun to participate in shrimp production, they see no future in it, either for their families or for their communities. This transition has been associated with a loss of agrarian livelihoods in the form of rapidly declining opportunities for sharecropping and agricultural labor, the disappearance of access to resources that once supported subsistence, and the ongoing dispossession of local residents (Paprocki and Cons 2014). The disappearance of rice is continually invoked to express the experience of this ruination of livelihoods. As Arjav did, residents discuss the disappearance of rice both as *dhan* and as *bhat*, meaning rice in its cereal form (raw or in the field) and in its cooked form for eating. In this way, they lament the ruination of agrarian modes of both production and social reproduction. As the agricultural crop at the heart of agrarian economies and production systems throughout the region, rice has sustained these communities for centuries. As the staple food of their diets, rice is produced for subsistence as well as for selling, and it has nourished their families and provided for their survival in the best and worst economic times. In both senses, then, the continuation of shrimp aquaculture threatens rice in the village and its contribution to the continued survival of the village and its inhabitants.

The contradictions between shrimp cultivation and rice farming are at the heart of the social and ecological transitions taking place in Khulna. Though agrarian dispossession in rural Bangladesh has been an ongoing process since at least the colonial period, shrimp aquaculture has entailed a pivotal rupture of agrarian political economies in communities where it is practiced. For this reason, I intentionally avoid referring to shrimp aquaculture as "farming," which for village residents denotes a particular sociocultural structure to which the cultivation of shrimp presents a stark contrast. In this chapter, I examine this contradiction between rice and shrimp through attention to stories of residents about this loss of rice (in its multiple dimensions), and the implications for lives and livelihoods.

In this chapter, I examine the agrarian transitions taking place at the nexus of two particular dynamics in Khulna: first, the transition from rice agriculture to shrimp aquaculture for export; and second, the emergence of new visions of "developed" futures in the age of climate change. As stories from Kolanihat demonstrate, these particular visions of developed climate futures, which are central to the adaptation regime, are associated with the decline of existing ways of life of the current residents of the region. Besides existing in geographic proximity to one another in this region, shrimp aquaculture and climate change adaptation are closely linked to one another through the overlap of their respective social and ecological causes and effects.

I explore these agrarian transitions through three key empirical interventions: First, I examine the ways in which shrimp and rice cultivation are antithetical to one another and the implications of this tension to the political economy of agrarian change in this region in which they have collided. A transition to shrimp culture entails an extraordinary transformation of the coastal landscape and the dynamics of production that have historically sustained its populations. Second, insofar as shrimp production entails a radical reduction in labor requirements and thus in the number of people working in and inhabiting these communities, the result of the transition to shrimp aquaculture is the displacement of farmers and communities from this landscape. I explore this transition through an examination of narratives about it among those who remain, who have been impacted by the transition in diverse ways. Third, the resulting migrations out of the coastal zone reflect this process of depeasantization. I examine this migration through testimonies of people who have left Polder 23 and are now living in a slum in Kolkata populated by migrants from Khulna. I argue that the discourse concerning the impacts of climate change in Southwest Bangladesh, which has framed these migrants as "climate refugees," both obscures and facilitates these dynamics of agrarian transition. The chapter primarily draws on ethnographic fieldwork conducted in Kolanihat, a village in Khulna's Paikgachha *upazila*, in Polder 23. It also makes use of interview data gathered in a participatory study carried out in the area in 2013 using Community-Based Oral Testimony (more details on this study can be found in the Methodological Appendix).

Paikgachha is an important site for understanding the growth of the commercial shrimp sector in Bangladesh, both due to the proliferation of shrimp culture in the area, as well as to its role in the expansion of shrimp culture throughout the region. In 1968, consultants for USAID recommended that the government establish a shrimp culture research station in Paikgachha, explaining their site selection by citing the damage that the nascent polder system was already causing to

FIGURE 5.2. A billboard in a public market in Paikgachha advertising a program supported by USAID, WorldFish, and the Bangladeshi government to promote the use of shrimp fry that have been tested for the presence of viruses.

Photo by the author.

local agricultural production there (Swingle et al. 1969). The drainage and salinity problems generated by the polders would make Paikgachha the perfect site for experimenting with the opportunities afforded by this ecological crisis, consultants reasoned. In 1984, the World Bank approved funds for a National Brackishwater Research Station in Paikgachha, indicating that establishing such an institute there was necessary to take over from a research institute in neighboring Satkhira district, where shrimp culture had already resulted in salinity levels high enough that, they concluded, crop growth was no longer possible (World Bank 1984). In addition to having been identified as an important site for researching the potential of shrimp aquaculture, Paikgachha has also been home to a great variety of donor- and NGO-sponsored development programs aimed at expanding the reach of shrimp production, including major programs supported by USAID, the World Bank, and the Dutch government (see figure 5.2).

Kolanihat

Kolanihat is one of thirty-one villages in Polder 23 and is home to a population of about 700 people. It is located about five miles from Paikgachha town, a small market town of about 16,000 people (BBS 2016). Before the shrimp boom, many of the residents of the village were landless people and marginal farmers whose livelihoods depended on sharecropping and agricultural day labor. A survey from a similar rice-producing village in neighboring Polder 22 in 1987 reported that over 50 percent of households fell into this category of marginal laborers (Datta 1998, 31). Until the mid-1980s, most residents of Kolanihat report that they produced one or two agricultural crops per year, depending largely on the elevation of their particular plot of land. *Aman* (monsoon season) rice was historically the most important crop, with abundant yields resulting from the high fertility of the alluvial soil. Residents report that historically their *aman* yields were high enough that they could survive on the single crop throughout the year if their land was unsuited to more than the single monsoon growing season. This is confirmed by records from the colonial period that corroborate these claims of the historical importance of the *aman* yield in Khulna (Bengal Government 1898). In addition to *aman*, residents report having grown jute, sesame, mung beans, and a wide variety of other fruits and vegetables.[1]

This agriculture was made possible by irrigation from the river that runs south of the village, which delivered fresh (nonsaline) water for six months out of the year. During this time before shrimp production, they could extend their growing season with canals fed by the river that could be damned up to store freshwater and extend its availability further into the dry season (when the river water becomes more saline). Before the polders were built, the village also practiced a traditional form of water management known as *oshto masher badh* ("eight month embankments"), which was part of the historical village- and farm-level hydrological regime examined in chapter 1.[2] *Oshto masher badh* was a system used throughout much of the coastal region in which laborers would annually build up earthen embankments to protect agricultural lands from the river and subsequently tear them down to allow tidal inundation for the remaining four months, thus facilitating sedimentation and fertilization of the soil.[3] The system extended the growing season by protecting land from salinity and flooding, while also not being so permanent as to contain unwanted floodwater and prevent the mitigation of waterlogging. It required a high degree of cooperation between landlords and sharecroppers, as the latter generally undertook the embankment work as a condition of their tenancies. After Partition, when many Hindu landlords migrated to India, the system began to break down along with the cross-class coalitions among the peasantry.[4] Any remnants of the informal system that remained were finally ended after the construction of the polders, when the permanency of the concrete polder infrastructure precluded the manual construction of such ad hoc mud embankments.

FIGURE 5.3. Wakil's sluice gate.

Photo by the author.

In 1986, as Bangladesh's shrimp export boom was taking off, investors from Khulna started to arrive in Polder 23 as they were interested in establishing ghers on agricultural lands. In Kolanihat, the largest among these was Wakil Saheb ("Saheb" denotes his wealth, power, and high social status), a businessman from Khulna who today still owns a gher of approximately 400 acres in size.[5] Wakil's gher is managed by Saiful, another outsider, who is from the neighboring district of Shatkhira. Saiful lives in a large house in Paikgachha town, and he manages the guards (also outsiders), who stay in the village. Wakil himself is thus several steps removed from the actual work of shrimp cultivation in Kolanihat. His gher is situated in a long strip running parallel to the river. Wakil has also installed his own sluice gate to regulate the flow of water in and out of the village (see figure 5.3), which operates autonomously from the municipal sluice gate of the Bangladesh Water Development Board. Wakil's sluice gate is used to flood all the land in the village, primarily during the dry season, when salinity levels in the river are the highest and thus most hospitable to shrimp cultivation.

Accounts in Kolanihat differ about the amount of force used to compel this transition to shrimp production when Wakil arrived, as well as the amount of resistance that was proffered by local residents. Landless people described sentries hired by these outsiders either to force people to give up their land or to guard the ghers once they were established.[6] Accounts of the village's wealthier residents and landholders deviate in varying degrees from these accounts of overt force being used to compel the transition to shrimp. This ambiguity is likely due to different levels of complicity and consent at different points during the transition.

Of particular concerns to all residents, though articulated most strongly by the poorest, is a common sense that the outsiders who came to engage in shrimp production are guilty of harassing and committing violence against women. As one resident explained, "If it gets too late coming back from work, either cleaning moss or catching crabs from the shrimp ghers, you get harassed and insulted. And if a woman is found alone, she gets raped, so the roads aren't safe for women." Though I did not hear any specific accounts of overt sexual violence, the concerns speak to a broader tension between local residents and outsiders hired to do work in the ghers, whose task is to occupy and patrol a landscape that was formerly inhabited much more communally. Another resident explained,

> The situation is such that there is gher after gher and there is very little road to walk on and even if you do walk, each gher owner on each side will accuse you of stealing from their gher. Sometimes people cannot get from place to place because they do not allow travel on those roads. They will accuse you of spreading [shrimp] virus in the gher and causing them losses. I don't want to get caught up in these troubles, so I try to stay away. Some have to travel on those roads, though. Most stay away. Those who do [walk past the ghers] have sometimes been physically assaulted.

This sense of occupation is exacerbated by the great extent of land that has been taken over by the ghers. Indeed, there is only one narrow path through the village where one can walk that is not directly adjacent to the ghers; it winds between a canal and the thin strip of land on which most of the village's homesteads are now located. Thus, the feeling that the ghers have taken over the physical geography of the village, encroaching on the space not only for rice farming, but also for walking, gardening, playing, and otherwise inhabiting, exacerbates the sense among the village's residents that they are being pushed out by shrimp cultivation.

Now when smallholders recount the initial period of rapid and overwhelming transition, it is often inflected with a sense of regret and confusion. Many smallholders were offered contracts to buy or lease their land, though they report rarely or never having been paid, nor having had a choice in whether they would give their formal consent. One landless woman explained, "The rich people who do shrimp farming often don't even pay the rent for the land that belongs to poor people. They use their power to keep stalling, 'I'll give the rent later, tomorrow, don't bother me now.' They keep saying this. They cultivate shrimp using their might." In this sense, the cultivation of shrimp shifted the control over land and resources in the village as well as the dynamics of consent over the use and appropriation of property and the terms of production.

Some people reported having their land stolen through different manners of legal maneuvering, such as one man from the village next to Kolanihat, who explained,

> We had a lot of land, 18 *bigha* [6 acres]. But there were complications with that land. A man named [Debjit Prachanda] from the neighboring village drew up a fake document and we were forced to give him half of our land. Nine *bigha* [3 acres]. I had to give some from my share, my older brother had to give some from his share.

When residents describe this kind of legal mechanism of dispossession, the statement is often somewhat vague and suggestive of simultaneous frustration and shame. The legal benchmark for literacy in Bangladesh is one's ability to sign one's own name. If one signs a legal document that one is not able to read, there is little recourse to remedy the consequences of having been deceived about the document's contents. This particular shrimp industrialist, Debjit, is unlike Wakil in the sense that he is a neighbor, a member of the immediate community and someone with whom villagers must interface on a regular basis—in relation not only to questions about land titles and cultivation but also questions about any other issues that might arise in the village, such as arbitration in the village *shalish* (local court), the organization of local *pujas* (Hindu religious festivals), or any other community concerns. Though this was his natal home, Debjit had been a businessman living outside the village until shrimp cultivation took off in the 1980s. When the shrimp boom began, he moved back to the village and began amassing a large gher made up of the former rice fields of neighboring residents. Now Debjit lives in the tallest and newest *pucca* house in his village. A conspicuous three stories high, the house is made of cement painted an arresting shade of bright chartreuse, standing out amidst the surrounding gray mud bungalows. Even as residents voiced their frustration about losing their land, they also expressed caution about wanting to avoid an outright altercation with this very wealthy and powerful neighbor.

While several residents describe this kind of overt land grabbing, many also report that they signed leases to allow the use of their lands for shrimp, though they also explain that they did not want to sign the lease or that they felt that they had no recourse when the terms of the lease were subsequently violated. Gorongo, a local schoolteacher and self-proclaimed cricket star who was the owner of a 3-acre gher, chronicled this history for me one afternoon over tea:

> Lots of guys came from the outside and took leases then. Our environment was *so* beautiful. Everything was green. We loved it! But then they came and took all the land. They gave 500–700 taka per *bigha* [$18–25 per acre]. It wasn't our choice. They did so well. They built big, beautiful buildings in their own areas. But we didn't develop here. We didn't get any benefit.

As Gorongo explains, the question of "choice" over participation in the transition to shrimp is fraught with ambiguity for smallholders throughout the region. They express regret over their initial assent but anger over their inability to recover their land. Another resident of a neighboring village recounted his struggles over the terms of his lease:

> These people are the ones who allowed the salt water to be brought in and they are now suffering. They are not able to push out the outsiders or get out [of the shrimp business] themselves. It is the local people who made mistakes. They rented out their five *bighas* [1.7 acres] of land. Now I am in trouble because where I signed the deed for three years, he rewrote it to be twenty years. Mr. Wajed Ali came and made a lease for three years. . . . [7] Now when the three years elapsed, we did not want to renew it, but then he showed that the deed said twenty years. At that time, he was very powerful, you could say he was the government. . . . We do not understand too much about the law, but somehow he showed that we had signed a twenty-year lease, whereas when we signed the document we knew it was for three years. . . . He told us nice things and took our signatures on official paper and then wrote twenty years on the lease. What will you do? If you go to attack him, the police will come to take us away. They'll say, "Didn't you pay attention when you signed?" What can we say now? If we complain to the administration, they'll say, "You signed there for twenty years. If you hadn't signed, it wouldn't have been twenty years." They harassed us into this situation.

As this smallholder describes, he feels some sense of responsibility for having allowed the initial encroachment of shrimp into his village but also regret over the feeling that villagers had been tricked into a permanent loss. Importantly, he also articulates anger over the complicity between the local authorities and wealthy outsiders and a frustration that this elite collusion has entrenched the shrimp industry in their region.

Beyond these specific disputes over land tenure, conflicts over water management became the primary arena through which contestations between rice and shrimp played out. Hydrological units within the polders must be managed collectively, as water intake and drainage are shared across broad areas, usually the size of a full village or larger.[8] As such, control over the water regime is necessarily centralized, meaning that whatever authority controls the water of an area has outsized control over the agricultural production of all the individual plots within that area. With drainage access to the river obstructed for interior smallholders like Gorongo, the rest of the land beyond Wakil's gher remains waterlogged year-round. Thus, Wakil has the power not only to continue cultivating shrimp on the 400 acres over which he has maintained control, but also to make production decisions for the entire village. If Wakil wants to keep cultivating shrimp, then the village cannot return to rice farming, nor the social life and agrarian political economy associated with it.

This control over the hydrological system can be secured either through political means, or by surreptitiously breaching the embankments dividing plots of land. In many communities, this is also achieved by boring holes in the embankments to install PVC pipes for water intake as an alternative to a full sluice gate. When farmers in Kolanihat describe their resistance to the expansion of shrimp aquaculture, it is primarily these water management conflicts that they highlight. As one farmer whose land had been taken explained,

> The salt water [from the ghers] destroyed the roads and nobody fixed them. Slowly famine started to show. Business slowly started to break down. I would go to people and say, "I am not even able to eat rice; how will I pay you back?" I was born in the area, so I know that if I am unable to eat, how can I force someone else to pay me? . . . There has been fighting, cases have been filed [against] the businessmen who have ghers. There have been clashes with them. These people live in the city, some live in Khulna, Shatkhira. The rich people who control the administration have been torturing us. We seal the WAPDA [embankment][9] and then they go at night with police and they break it. When we go out in the morning, they send goons hired from the city to attack us. They torture us. If we go to the police station, they make us file a General Diary [police report] and they say, "We will look into it." They say they will look into it, but that very night the water is released into the gher again.

This testimony highlights both the physical force employed in compelling the transition to a saline landscape for shrimp cultivation, as well as the complicity of local government officials in exercising that force. Although Wakil operates his sluice gate independently, residents explain that this is done with the implicit support of local officials, who not only do not restrict the intake of the water through his sluice gate but also do not facilitate its drainage. Another resident explained,

> If you want to bring salt water into an area, you need permission from the local government to close the gate, to put in a channel. And if the rich people don't decide on this, then people who don't have much money or have no money can't do anything about it. If the rich people decide to do it, they need to go get permission from Paikgachha *thana* [police station] to open the gates so they can cultivate shrimp. It takes 50,000 taka [$590]. People like us can't go and get permission from the Paikgachha *thana*, we won't even know anyone there or know what to say.

This disproportionate control of elites over water infrastructure is magnified by the domination of their decisions within the entire hydrological unit. Even where an embankment is not intentionally breached between fields, a shrimp gher adjacent to a rice field causes water seepage and salt deposits, destroying the rice crop itself and causing increased levels of soil salinity. This prevents cultivation decisions from being made independently. Thus, control over the water management regime becomes de facto control over production decisions in the entire area.

These conditions of conflicting water management paradigms have been foundational to the expansion of shrimp aquaculture throughout Khulna. In 1993, Adnan wrote that early analyses by development programs of the economic benefits of shrimp cultivation in Khulna were based on assumptions about effective and equitable water management, allowing for the cultivation of both rice and shrimp (1993). However, he explained, such assumptions were flawed due to these inequitable water management practices, "Efficient management of water flows is only possible for those who own or operate the appropriate water control mechanisms—typically the large shrimp-growing business interests, rather than the poor or middle peasants whose lands have been inundated with saline water against their will, leading to involuntary shrimp culture" (1993, 2). This involuntary shrimp culture is the circumstance in which smallholders in Kolanihat and elsewhere in Khulna now find themselves. It is the condition, in varying degrees, reported to me by every smallholder gher operator with whom I spoke while in Kolanihat.

Landscape Changes

These transformations in water and production regimes have been accompanied by similarly dramatic transformations in the local ecology. Among these changes, villagers cite the death of most of the village's trees most frequently as evidence of the transformed landscape. They lament in particular the loss of the fruit trees, noting the former presence of coconut, date palm, tamarind, banana, guava, mango, jackfruit, pomegranate, wood apple, jujube (*kul*), and black plum (*jaam*) trees. Since the time when shrimp cultivation began, these trees all died due to the soil salinity. The few trees that remain now are *keora* trees, a mangrove species that can survive in moderate to high soil salinity, which is considered more tolerant to growth in highly degraded landscapes though it doesn't contribute to subsistence in the form of fruits or fuel wood.

The loss of these fruit trees has contributed to an overall decline in subsistence production capacity in the village. The loss of farmers' ability to grow and store rice for family consumption has had the most serious impact on subsistence. However, the ecological collapse that precipitated the death of the trees has had a range of other impacts that have shaped the ability of villagers to survive in this landscape. As the ghers expanded, moving closer and closer to residential land, the salinity seeped into the soil surrounding their homesteads, affecting the plots where they used to cultivate gardens. These gardens formerly supplied an abundance of vegetables that fed the families of the landless and land-rich alike throughout the year. Many people reported that at that time not only were they able to grow enough to feed their families, there was such an abundance that they would share these yields from their gardens with their neighbors. "It was not possible to eat all of them," one man said. "We didn't sell them; whenever someone in the area needed a particular vegetable, we would give it to them. . . . Now it has become salt water. I myself have to buy all my vegetables, so how could we give to others?" Villagers mentioned cultivating okra, eggplants, pumpkin, string beans, radishes, potatoes, taro, and several different varieties of greens and gourds, none of which will grow now.

Kolanihat residents explain that the salinity seeping into the soil has also infiltrated the drinking water supply. One staff member in Dhaka managing a World-Fish project supporting shrimp cultivation in Paikgachha said to me, "Some people said, 'Our drinking water is salty because of shrimp,' but I don't know how you can really prove that." Indeed, as with many other aspects of ecological change in Khulna, "proof" of what is driving up the salinity in drinking water in Paikgachha is elusive. People in Kolanihat trace the salination of their drinking water to the transition to shrimp cultivation. Previously, the primary source of fresh water in the village was ponds that were recharged with rainwater annually during the monsoon. The water from these ponds was used for a variety of household purposes—bathing, washing clothes and dishes, and drinking and cooking (they would boil the water before drinking it). These ponds used to be situated near people's homesteads and would usually be accessed communally by neighbors from the surrounding area. Now that the domestic space has dwindled, contracting as the ghers have expanded, there are no more ponds in the village. There are also some tube wells from which people used to retrieve drinking water, but people say the quality is no longer good, as it has become saline and is otherwise contaminated, making it unsuitable for drinking.[10] They also harvest rainwater for household use, and some have devised systems of running ropes from troughs on the roofs of their homes into large earthen drums on the ground to collect the water. However, without larger receptacles for the water like the ponds, these home-based tanks usually catch enough water to last only a couple months.

As a result, villagers are now forced to purchase virtually all their drinking water, which women and children carry in from outside. Some travel to a neighboring village about 10 miles away, where there is a tube well that supplies freshwater, and some go to a pond in Paikgachha town. They pay 10 taka (about 12 cents) to fill a 10-liter aluminum vessel. Most members of the community travel by foot to carry water back, while those who can afford it pay an additional 10 taka to travel by *nosimon*, a kind of motorized rickshaw van that plies the major road through the polder carrying passengers and cargo. Most families report that they generally need more than one of these 10-liter vessels to meet their domestic needs in a single day.

In addition to this water, villagers are also now forced to purchase cooking fuel from Paikgachha. Before shrimp cultivation, there were large areas in the village where cows could graze on grass. Afterward, however, these areas ceased to exist, as the communal spaces gave way to ghers and salinity killed the grass on the land that remained. In the absence of grass, the only way to raise cattle is to purchase hay from Paikgachha. In addition to providing milk, these cattle also supplied residents with the majority of their cooking fuel. The traditional fuel, called *ghute*, is made by gathering cow dung and patting it into round cakes or molding it around long sticks, which are then dried in the sun. Now that there are few cattle left in the village, residents purchase fuel from Paikgachha town for 160 taka per bunch (about $2), plus 10 taka (12 cents) for transport by *nosimon*.[11]

The enclosure of these commons also extended to a large canal running through the village that used to serve as the primary outlet for drainage and inlet for fresh irrigation water.[12] In order to retain the water for shrimp cultivation, gher operators have blocked the canal and turned it into another space for shrimp cultivation. As one woman explained,

> All the water bodies belong to someone [now], they won't let you catch fish from there. Before, all the rivers, wetlands, and lowlands were open for everyone to go fishing. The rich, the poor could all catch fish and eat them. No one would stop them. . . . The rich people have paid money

and taken land from the poor. . . . All around is just water and water; we don't have the beautiful, communal, and wholesome environment that we had before.

As this resident suggests, when the canals flowed directly from the river, they were full of a great diversity of wild fish that residents could capture for their own consumption (residents mentioned several native freshwater species, including *rui, shoul, baila, boaal, taki,* tilapia, carp, catfish, climbing perch, freshwater prawn, and crabs). Now that the canals have been closed off, access to fish has declined and many people report that they are unable to afford to purchase them to feed their families. Some allow tilapia fish to grow in their ghers along with shrimp; they call these *khaoar mach* ("eating fish") because they cannot sell them so there is sometimes access to these for local consumption. Several expressed frustration with this dwindling diversity, while one woman said, "How many times can you eat the same tilapia? . . . I am fed up with having this tilapia, I don't even like it anymore." Several other residents also expressed dissatisfaction with an ironic state of affairs in which the expanded production of fish for export in their area resulted in a scarcity of fish for their own consumption (the Bengali word for fish, *mach,* is also used to refer to saltwater shrimp, which are called *bagda mach*). Another woman explained, "We can't always get fish. It's like being thirsty when you're in the middle of the ocean surrounded by water. We are surrounded by fish here, but we're always craving it." These active dispossessions combine into an assault on access to almost every means of subsistence previously available to villagers in this community.

Tenure Transitions

For about ten years, the land tenure situation in Kolanihat went on in the same condition as when shrimp cultivation started, with Wakil's massive gher covering most of the village. As time went on, the smallholders whose land had been taken over received fewer payments or rents even ceased altogether, and those who had initially been open to the arrangement became increasingly disillusioned with it. Then, in the mid-1990s, these land tenure conditions started to shift. There is a very large, but dilapidated, colonial-era *zamindar bari*[13] on a spacious, solitary plot of land on the opposite side of the canal from the rest of the village's residential area. Today this *zamindar bari* is inhabited by Radhika, an elderly woman whose father-in-law lived there before Partition, along with four sons (who also have other residences outside the village), daughters in law, and several grandchildren. One of Radhika's sons is a doctor and another is a member of the Union

Parishad (local government) committee, which makes him a very powerful member of the community. The family sleeps on wooden cots on the ground floor of this three-story mansion, and their plastic furniture and other modest and sparse fixtures suggest some deterioration from what was once presumably a lifestyle of great affluence and comfort.[14]

Radhika told me that when Wakil first came to Kolanihat, her family was enticed by the promise of a lucrative lease and agreed to rent out their large plot of 67 acres to Wakil to incorporate into his gher. Like other large landlords in the village, they had previously leased out most of this land to local sharecroppers for rice farming. This assent of the wealthy local landholders gave Wakil a great deal of legitimacy and power in the village and made it more difficult for those who owned less than an acre to resist the loss of control over their own land. After ten years, however, Radhika's family was only receiving 4,500 taka ($50) per acre from Wakil, and they had become disillusioned by the deteriorating economic conditions and their limited share in the considerable profits. Radhika told me she thought the rent should have been 30,000 taka ($350) per acre (though most smallholders told me that the accepted rate for most people in the village at the beginning was 1,500 taka [$18] per acre, and now it is between 12–21,000 taka [$140–250] per acre). When Wakil refused to return their 67 acres to them, Radhika's family pressed charges against him with the local police. Their family's wealth and political position in the community gave them significant influence with the local authorities and their case was successful.

Once they regained control of the land, what remained of Wakil's gher still stood between their plot and the river, making access to water and drainage difficult. Radhika told me that her family did not want to continue allowing the gher on their land, but they did not want to give up their land either, and under the circumstances, they had no choice. Also, having observed Wakil's success, they thought they might get some benefit from having their own shrimp gher, and so they began to cultivate shrimp independently. When I asked Radhika how they like the shrimp gher business, she looked at me impassively and said, shrugging, that they did not really care as they hired other people to do the work. On another occasion, she told me that they were disappointed with the profits they earned from shrimp production, so they did not really like it. This ambiguous indifference may or may not be genuine; several smallholders reported to me that Radhika's family support the continued production of shrimp in the village and that they were among the greatest obstacles to a collective mobilization aimed at transitioning back to rice farming. In this sense, despite their earlier conflicts with Wakil, Radhika's family may now be allied with him (either implicitly or explicitly) in ensuring the perpetuation of shrimp production in the village.

In addition to their shrimp gher, however, Radhika's family has 1 acre of land adjacent to their home where they have continued to successfully grow a crop of *aman* rice. They have their own private tube well that they can use to supplement rainwater for irrigating a garden (and from which they also retrieve drinking and cooking water for their own household use). The land is elevated near their home, being at a higher point than the surrounding ghers, so salinity seepage poses less of a problem than elsewhere in the village as gravity facilitates drainage. To cultivate this land, they usually employ about seven people in a single rice growing season. For their 200 acre gher, by contrast, Radhika said two of her sons look after managing its operations (in addition to the work they do in Paikgachha), and once or twice a month they hire a group of five to seven women to do *sheola kaj*, (collect the scum off the surface of the water, a task that I explore in more detail later in this chapter). Before the shrimp boom, they said, they had cultivated both *aman* rice and sesame, a winter crop that requires very little water and is relatively tolerant of salinity, making it well suited for cultivation in the saline-prone parts of the coastal region. The people they hire for this work of rice cultivation are local laborers, as this is one of the only agricultural opportunities left in the village. An elderly couple doing mid-season weeding work in the rice field surrounding Radhika's *zamindar bari* told me that they were formerly sharecroppers but that once the land they used to farm was leased out to Wakil, the labor in this remaining rice field was the only livelihood left for them in the village.

Though Radhika says their rice yields are slowly declining every year, they still get about one third of the yield they would under normal conditions. Radhika and her family keep all the rice grown on their land for their own consumption, and despite the declining yields, they say it feeds them throughout the year. Yet besides the income from the gher, it is not their only source of livelihood. Radhika's sons do business in Paikgachha, and they use this income to supplement their consumption. They thus inhabit the same ambiguous occupational category as Arjav's sons; although they earn some income from shrimp, it is their off-farm earnings that sustain their family.

Once Radhika's family had regained control over their own land, a precedent was set that created an opening for the smallholders who remained in the village. One small gher operator described this to me as "*gono dokhol*," which means to collectively take possession or control. Yet the recovery has been uneven, and those whose adjacent plot holdings are smaller or are surrounded by Wakil's still-sizable gher have been unable to reclaim their land. As Gorongo, the cricket star, explained to me,

> So, we saw that they did so well, we wanted to try ourselves. We had a struggle with them, we went to war. And through the struggle, we got

> the land back. . . . Some landless got work in shrimp trading, but very few. Everyone else had to leave. It's very important that you tell people, like at big seminars, that we want our land back. You have to take this story back with you and tell people. Many people here who just had small plots of land have never been able to get them back.

Gorongo's statement here is indicative of the descriptions of this period of land repossession among smallholders. They articulate the hardships they experienced in losing access to their land, then express the struggle to have it returned to them, yet they also appear hesitant to characterize this recovery of their land with any sense of triumphalism. As they describe their current conditions, it becomes apparent that this hesitance is primarily due to their sense that they still do not have autonomous control over their land and they have been largely unsuccessful in managing their own ghers. Even as Wakil technically returned their land to many of the remaining smallholders, with drainage blocked and all of the land in the village perpetually covered in salt water, their choices are severely constrained in how they can use it. Instead of recounting their success in regaining control of their land, they recount the struggles they experience now in trying to earn a livelihood from it.

Today Gorongo operates a gher 3 acres in size. In relation to average landholdings in rural Bangladesh this is a very large amount of land, while it is among the smallest shrimp ghers. His results have been mixed. Sometimes the profits can be good, but the threat of diseases that plague the shrimp industry here (in particular, white spot syndrome, which villagers refer to simply as "virus"), poses a constant threat of killing all the shrimp in a gher as it spreads quickly between plots. Because shrimp production requires a significant up-front capital expenditure each season, the loss of a gher full of shrimp to the virus can be catastrophic. The fry that populate a gher must be purchased either from dealers who get them from a hatchery or from people who collect them wild in the river. Operators also usually purchase other inputs such as feed and fertilizer. Each of these items is available in variable qualities and costs, although development agencies say that using an inferior quality of fry and other inputs results in both lower yields and greater susceptibility to disease. The international shrimp market also fluctuates significantly, which has a dramatic effect on these smallholders. In the wake of the global economic meltdown, their returns plummeted below the break-even point despite a national supply shortage that should have driven prices up (Moni 2014). These declining rates are likely due both to the power of the many middlemen within the country's lengthy shrimp supply chain and to a sharp and unexpected plummet in exports to the United States, where importers were prioritizing cheaper and lower-quality imports from China, Vietnam, and Thailand (Haque 2014).[15]

In order to afford the significant up-front investments required, smallholders like Gorongo take out microcredit loans at the beginning of the season. When their gher is struck with the virus and all their shrimp die or they do not earn enough to break even, they take out more loans to repay the previous ones. I regularly heard that there were many people who had lost their land in this way—after years of trying to get their land back, they finally regained control only to lose it altogether when they could not afford to manage the gher and all the costs that entailed. These people, who have lost their land through debts to microcredit agencies or wealthier neighbors, have all left the village.[16] The staff at a local BRAC microcredit branch office told me that most of the loans they give in Paikgachha are for investments in ghers and that they encourage borrowers to take loans for shrimp cultivation because they believe it is good investment in the area. One collection agent said to me, "When you look around this area, you can see that it is more developed." When I asked for clarification of what made it more developed, he looked at me as if it was obvious—"because of all the ghers." By continuing to encourage the use of microcredit loans for investments in shrimp aquaculture, microcredit agencies participate in the adaptation regime's promotion of this particular vision of development for rural communities in Khulna.

In this context, despite intermittent success with his own gher when prices are high and the virus is kept at bay, Gorongo explained to me that he likes farming rice and would love to return to it, but with big ghers like Wakil's, the choice is out of his hands. "If everyone was doing rice, then we would be happy," he said. "We all want to farm rice, but we can't. . . . If everyone moved back to rice, we would be free of saltwater. But without that, we can't do it." This antipathy toward shrimp cultivation is the prevailing sentiment among smallholders in Kolanihat. In casual conversations, sitting together in small courtyards and tea stalls, questions about shrimp provoke deep groans and eye rolls as residents throw up their hands in resignation and gesture toward the surrounding barren landscape. They explain that they sometimes earn well when the price of shrimp on the international market is high, but these markets are erratic, and therefore livelihoods dependent on them are unreliable. The word most often invoked to describe their conditions under shrimp cultivation is *obhab*, meaning scarcity or deficiency. One woman, whose husband manages a 3-acre gher, complained about Wakil blocking the river and explained, "We have scarcity because we don't do rice."

Besides Radhika's family, a few others have occasionally attempted to farm rice in the village under the current water management regime. One man who now lives in Paikgachha town was able to lease a 2-acre plot of land from some wealthier landholders in Kolanihat around the year 2000; the land is located near the river, outside the contained area that is flooded by Wakil's sluice gate. Prior to taking this lease, he had been working as a shrimp trader, but he said he really

FIGURE 5.4. Rice plot near the river recently converted back from shrimp *gher*.
Photo by the author.

wanted to go back to rice farming. Although the land had previously been used as a gher, he decided to try to flush the salt and use it to cultivate an *aman* rice crop, which he has been doing over the past several years. He told me that every year he has been operating at a loss with the rice, with low yields and significant costs for seeds, insecticides, and renting the land. The yields were increasing, however, and he said he believed that eventually enough of the residual salt would be leached that he could turn this rice plot into a successful operation. It is evident in seeing the crop that the recovery of the soil is still ongoing but that cultivation will not be impossible (figure 5.4 shows the crop in mid-season). For now, he and his family are just eating the rice, but he thinks that within a few years he will have a surplus to start selling. His growing success with rice cultivation demonstrates its continued potential in this area, despite claims that these shrimp-producing landscapes are no longer viable for agriculture.

I also observed a few other people inside the embankment in Kolanihat and the villages around it who were attempting to farm rice during the *aman* season. In Kolanihat they were concentrated on the far side of the village from Wakil's gher, and I could see that they had fashioned small drainage ditches to try to siphon off the saltwater. While no one had done this for any significant period of time, some told me that they had been doing it for a few years despite meager yields. They said that their farming was hampered by insufficient drainage and residual salt deposits in the soil but that the monsoon brought enough freshwater to flood the rice paddies. They estimated yields between 900 and 1,300 kg per acre, which is about one half to one third of standard rice yields (one person cited yields as low as 220–330 kg per acre). The best they could hope for would be to break even, a couple of them told me. Given these low yields and bleak prospects, I asked them why they continued to try to cultivate rice. They said that they had ghers on their land in the dry season but they were not having much success with either shrimp or rice, so they had decided that they might as well try farming rice, since that way at least their families would have something they could eat. It seemed like a better gamble, given that they could lose everything if their shrimp were struck with a virus. One man who said that he expected enough rice to feed his family for five months told me that rice was better for the environment, so they hoped to keep expanding the area under rice cultivation in the village. These attempts, moreover, would help to leach some of the salinity buildup in the soil each season. Managing the salinity, the villagers explained, would hopefully prepare them for a quicker return to year-round agriculture if someday the entire village was able to transition away from shrimp. Despite the great costs entailed in producing rice under the current conditions, these villagers' commitment to continue their efforts reflects an optimistic vision of the future of these agrarian communities, at odds with that of the adaptation regime. In anticipating the possibility of an agrarian future, they reject the ruination of the adaptation regime's dystopic imaginaries. In chapter 6, I examine village-level transitions in neighboring polders that are grounded in similarly optimistic imaginaries of the future.

Depeasantization

These attempts to continue farming rice also speak to a desire to preserve an agrarian lifestyle that is in the process of being lost.[17] People like Radhika's sons, who earn most of their money through off-farm livelihoods, no longer identify themselves as farmers. Even when they are speaking of their work with the gher within the village, they say "*chingri byabsha kori*" ("We do shrimp business") rather than "*chingri chash kori*" ("We farm shrimp"). Some smallholders say, "I am a farmer" but "I run a gher," as opposed to "I farm shrimp" (though the two statements are technically interchangeable). These shifting identities speak to broader transformations in the agrarian political economy of the village. Villagers in Kolanihat repeatedly spoke of the demise of the farmers in their community. As one former farmer told me, "Shrimp has destroyed all of the farmers." This is meant both literally and figuratively. Indeed, in the literal sense, many people have been forced to leave the village, and particularly landless people, for whom labor opportunities in the village have been eliminated. Remaining residents describe a great exodus in the aftermath of the shrimp boom; as one smallholder explained to me, "Those who didn't have land, they have all left." With opportunities for work disappearing, and their ecology no longer able to support the kind of subsistence production that formerly sustained them, life for landless people in these shrimp-producing landscapes is becoming increasingly untenable. A professor at a university in Dhaka with several decades of experience studying Khulna's coastal ecologies attributed this migration specifically to the transition from rice to shrimp, explaining that where saline shrimp are produced, "the area becomes almost uninhabitable," and that then "people have nowhere to go," thus indicating both the ecological and socioeconomic impediments to maintaining human settlements under such inhospitable conditions.

In addition to these literal references to the departure of so many residents, the demise of farming also refers figuratively to the people who have stayed but who can no longer be identified as farmers in the traditional sense. For some it reflects their profound struggles to survive in this landscape. One widow in Kolanihat who formerly farmed rice on a plot less than .2 acres in size (which has now been completely consumed by Wakil's gher) lamented the loss of opportunities for survival in the village, the lack of drinking water, and the failure of her garden. "We don't have any money; we can't pay for medicine or meet any of our needs. So, we are dying," she said to me. For others, like Arjav, it means that their survival has become so dependent on off-farm incomes that their identity as "farmers" has become compromised. Those who have stayed and today operate small shrimp ghers do so primarily through the support of off-farm employment by at least one if not several family members. These nonagrarian incomes increasingly become the backbone of, rather than a supplement to, this rural economy. This shift is the reason why they say their village will have no future unless they can find a way to return to agriculture.

Job Loss

Shrimp cultivation entails a radical reduction in labor demand relative to rice farming. While the shrimp boom has produced some labor opportunities in Kolanihat, the number of rural labor opportunities that have been created is dwarfed by the number that have been eliminated. As one woman explained,

> Previously, if one person had one *bigha* [one third of an acre] of land, twenty people would be employed for working on that land. They would

be busy cutting the earth, harvesting rice, plucking grains and various other activities. Those who did not have any land would survive just by working in the fields. Nobody was unemployed. Since the ghers happened almost everyone is unemployed. Nobody can get any work. A person with a gher manages it on their own. Only one person is needed to manage a gher. Previously one person would employ a minimum of five people but now more and more people have become unemployed. Now whoever has one *bigha* of land does not employ anyone else. They do the work on their own.

The result of this transformation is rampant unemployment among people who formerly depended on agricultural labor to survive. Joining their ranks are the landless who formerly depended on sharecropping, which is regarded as a more secure livelihood than working as a day laborer. Larger landholders like Arjav, who previously had arranged with landless neighbors to grow rice as sharecroppers on his surplus land, now no longer required such arrangements, and the availability of land for sharecropping quickly diminished. While rice is often grown on plots as small as one third of an acre, shrimp ghers operate at much higher economies of scale, as most gher operators describe plots of even a few acres as being untenably small. Most of the landless people, who have become redundant to this new rural economy, have by now left Kolanihat.

It is common for development agencies promoting shrimp aquaculture to claim among its economic benefits the creation of new jobs. However, most jobs created by the shrimp industry are in cities like Khulna, where laborers (mostly women) work in processing factories peeling and de-heading shrimp for freezing and packaging before they are exported. The gendered division of labor within this workforce entails significantly lower wages for the jobs dominated by women, as well as lower wages for women doing the same work as men (Islam 2008). In 2012, an analysis of the labor conditions for workers in shrimp-processing facilities in Khulna found that for a family of four, even if two earning members of a household were earning the legal minimum wage (which most do not), their earnings would be less than a living wage sufficient for basic caloric intake, noting that on average the families of workers in this industry consume 55 percent fewer calories than are required for a balanced diet (SAFE 2012).

The levels of labor demand within villages that have transitioned to shrimp are also a source of great uncertainty and disagreement. Nijera Kori organizers in Kolanihat and elsewhere in Khulna believe that shrimp cultivation requires about 10 percent of the amount of labor that rice farming requires; a member of the ESPA Deltas team (see chapter 3) told me that their surveys suggest roughly the same figure. Local residents cited figures as low as 1 percent, indicating the

TABLE 5.1 Calculations from various studies of labor requirements for rice and shrimp cultivation (in person-days per hectare per year)

RICE	SHRIMP
182.73 PD/ha[1] (*single cropping*)	80 PD/ha[2]
240 PD/ha[3] (*double cropping*)	80.4 PD/ha[4]
308 PD/ha[5] (*double cropping*)	106 PD/ha[6]

Sources:
[1] This figure is based on data collected in a primarily single-cropped area in Paikgachha in 1987–88 by Datta (1998), 43.
[2] Joffre et al. (2010), 57.
[3] Dey et al. (2012), 15.
[4] Alauddin and Hamid (1999), 291.
[5] Karim et al. (2006), 36.
[6] Nuruzzaman (2006), 448.
Note: PD/ha = person-days per hectare.

sense of significant loss of livelihoods experienced within local communities due to the rise of shrimp production.

Table 5.1 lists a series of calculations of labor demand for rice farming and shrimp cultivation from several different studies. The figures demonstrate a great deal of variance in estimates of the amount of labor required for both rice and shrimp production, although all figures denote significant discrepancies between rice and shrimp. Nevertheless, these figures also reflect greater labor requirements for shrimp than the estimates I heard from community members in Khulna.

Estimates from gher owners in Kolanihat also offer a different picture of these labor requirements, and while they do not suggest a clear pattern, they do give us a better sense of how the lower estimates in table 5.1 could have been calculated. The figures from Kolanihat listed in table 5.2 are based on conversations I had with six different gher owners or their agents who were based in the village. They are very rough because the shrimp cultivation season is not predictable. Two of the owners said they did not hire anyone because their sons did all the work (although they would occasionally hire groups of women for work in the gher once or twice a month). The estimates involving multiple laborers may be too high because there is rarely paid work available every day for large teams during the period when shrimp is being cultivated, so it is also likely in this case that many of these laborers were hired for only part-time work during these periods. All but the one man hired to manage the 10.8 ha gher said that if the shrimp were attacked with the virus there would be no need for them to hire any labor, and the discrepancies in time periods reflect this uncertainty about the survival of the shrimp. All these land plots are large enough that before the transition to shrimp

TABLE 5.2 Kolanihat gher owners' estimates of the amount of labor hired annually for shrimp cultivation

BIGHAS OWNED	LAND IN HECTARES	PEOPLE HIRED	APPROXIMATE PERSON-DAYS HIRED PER HECTARE (PD/HA)
200	26.99	5–8 for 1 or 2 months	5.6—17.8 PD/ha[1]
25	3.37	(none)	0[2]
170	22.94	2–3 for 5–6 months	13.1—23.5 PD/ha
18	2.43	3 for 1–2 months	37—74.1 PD/ha
7	.94	(none)	0[3]
80	10.8	1 man, year-round	33.8 PD/ha

Notes:
[1] Calculated at wages paid every day for 1–2 months, which is unlikely.
[2] Sons do shrimp work in addition to their off-farm labor.
[3] Sons do shrimp work in addition to their off-farm labor.

production they would have hired labor to support rice planting and harvest, and in the bigger plots they almost certainly would have leased out portions of the land to sharecroppers (which several of them confirmed to me).

This somewhat inscrutable collection of figures gives a sense of how unpredictable the availability of labor opportunities is in shrimp production systems; it also conveys that we should treat all figures relating to the labor in these systems with some caution. While some of the figures confirm the very low estimates of shifting labor demand provided by activists and landless people, the more important point to be taken away from them is how uncertain the work is for residents who would rely on it. As Edelman argues in relation to large-scale land grabs, focusing on the quantitative dimensions of this transformation may divert focus from the very real transformations taking place in social relations and livelihood patterns (2013). While support for shrimp aquaculture among government and development agencies has been justified on the grounds that it creates jobs in rural communities, these data suggest not only that the expansion of shrimp in fact decreases labor opportunities, but also that a focus on this quantitative metric conceals more than it illuminates about the impact of shrimp production on rural communities.

We must be attentive to these analytical discrepancies and what they reveal about the epistemological politics of research on shrimp aquaculture development. Both the figures cited by community members concerning labor market shifts as well as their interpretation of these figures are substantively different from those cited by recent studies carried out by aquaculture development agencies. How should we make sense of these inconsistencies? In his study of the debates between aquaculture development experts and the NGOs and activists who oppose the expansion of shrimp aquaculture, Béné describes how aquaculture "experts" discount the claims of the latter as misinformed, "incorrect drivel" (2005, 595). Like Béné, my analysis suggests that competing claims regarding the impacts of aquaculture in Paikgachha indicate that there is more to these discrepancies than faulty data. Rather, they may reflect different analytical foci and values with which differently situated researchers and community members approach the analysis.

One researcher studying shrimp and rural livelihoods in Paikgachha for an NGO in Dhaka told me that their survey research revealed that shrimp cultivation had caused residents' incomes to grow, thus causing an increase in food security. However, in focus groups, they found that people consistently reported that they were food insecure and that their food security had declined significantly from the period when they produced rice. The organization found the claims made in these focus groups to be dubious, given their own quantitative data indicating the contrary. The discrepancy between this NGO's data and the local perceptions may indicate a straightforward preference among residents for more stable consumption. Indeed, if the incomes of some increase considerably during certain parts of the year yet are not stable or predictable, and if subsistence options in the village have declined throughout the year, then they may experience the transformation as an increase in periods of food insecurity, despite an overall increase in cash flow. However, my research in Kolanihat suggests that the foundation of these perceptions may reflect a wider scope of values and concerns among community members than those related strictly to cash flow and income smoothing. Edelman writes that "every dataset has an implicit epistemology behind it" (2013, 494). This is certainly true of the figures in both tables 5.1 and 5.2, as well as my ethnographic data from Khulna. The testimonies presented in this chapter demonstrate that, for local residents, assessing whether there are sufficient, meaningful, and properly remunerated labor opportunities must involve much more than a quantitative calculation of cash flow and the number of jobs created or lost.

Shrimp Labor

A closer look at the type of labor available in shrimp ghers, however limited, facilitates a more nuanced understanding of how these assessments among residents are made. The most common work in the shrimp ghers is known as *sheola kaj* (literally "algae work"), which requires wading through the waist-high waters of the ghers using one's hands to skim the algal blooms that have collected on the water's surface and piling them into mounds every few yards (see figures 5.5 and 5.6). The work is conducted almost exclusively by women, who describe it as extremely unpleasant. Depending on the size of the gher, this work may be done from every couple weeks to once a month or less. It is carried out by groups of women, usually of about five to eight, starting in the morning around 8 a.m. and working until 2 p.m. They do the work with their saris tied up between their thighs, but the saltwater eats away at them, and one woman told me that she needs to buy four saris a year now instead of two because the fabric begins to fall apart so quickly due to the salt and chemicals in the water.

Women in Kolanihat are paid between 50 and 80 taka (60 to 90 cents) per day for their work in the ghers, usually for a few days a month or less. This is significantly lower than what men are paid for agricultural labor (100 to 500 taka or $1.20–5.90 per day, depending on the work and time of the year). The limited availability of this work in the ghers combined with the abundant supply of underemployed laborers also significantly depresses wages. One woman explained,

> The gher owners are very comfortable. The big ghers, if they have a good shrimp production one year, that money will last them for five years. The poor people, those who have to work day to day to earn a living, if they make even the slightest mistake they are kicked out and told they will be replaced by someone else immediately and that they are no longer needed. The owner makes sure that the new people that are brought in are paid even less than what the previous person was paid. Those people have no option but to do the work at the cheaper rate as they have nowhere else to go. They cannot even argue with the gher owner. For the gher owners, if they have a problem with the shrimp production one year, the next year the production is ok. He has money saved up so he can afford to buy shrimp fry and release it in the gher. What can a poor person do if he is forced to do 100 taka [$1.20] worth of work for only 50 taka [$.60]? He has no choice.

Several laborers also reported having their wages withheld for significant periods of time, even as long as a year, particularly when the market price of shrimp is low or the gher is struck with a virus and the gher owners' income is reduced. The precarious nature of this work is a cause of great concern for residents who have few other income-earning opportunities on which to rely. In Salabunia, another village in Polder 23 that has experienced a similar transition to shrimp (with associated employment and land tenure dynamics), one study found that only 2 percent of the jobs available in aquaculture in the village are permanent positions, with 61 percent of laborers working in temporary day labor arrangements in *sheola* and *mathir kaj* (a description follows) (Belton 2016). This contrasts with agricultural positions, which are usually contracted on a seasonal basis.

Once, while standing at the edge of a gher with a woman who was hired occasionally for sheola kaj there, as we both grimaced slightly while looking at the fetid

FIGURE 5.5. *Gher* covered in algae.

Photo by the author.

water and the foamy scum collecting at the edges, I asked her if she liked working in the gher. She looked at me with a guffaw, eyes wide in disbelief, and asked me, "Would you like to get into this water?" I looked down, struggling to make eye contact with her and feeling embarrassed to have asked what she obviously thought was a ridiculous question. The problem, she continued, is that it is only women who do this kind of work in the ghers, not men. As a result, the challenges of the work, and particularly the associated health problems, have received little attention.

Though women are understandably hesitant to open up about these concerns, some describe skin rashes and gynecological conditions that they commonly develop after working in the ghers. These concerns were confirmed to me by a doctor who runs the local government hospital. Upon walking into this hospital, one is greeted by a large vinyl banner buoyantly describing a pharmaceutical injection called "Pradox," with the tagline "Reactivator for Life" (see figure 5.7). The active chemical compound in Pradox, pralidoxime chloride, is used as an antidote to organophosphate poisoning, including chemical weapon nerve agent attacks against military personnel. In Paikgachha, the doctor explained, Pradox is administered to patients, primarily women, who have attempted suicide by ingesting chemical fertilizers used in both aquaculture and agriculture. He estimated that

FIGURE 5.6. *Gher* in the process of being cleaned, with some algae piled in mounds.

Photo by the author.

he sees approximately twenty such patients each month—women who have attempted suicide by drinking chemicals used for aquaculture. Pradox is used to resuscitate them. The doctor brushed off my questions about these apparently widespread suicide attempts dismissively, explaining to me that women often become hysterical if their husbands refuse to buy them a piece of jewelry that they want.

As he gave me a tour of his hospital, the corridors of which were crowded with recovering patients huddled on mats laid out across the cement floors, he listed the variety of ailments that afflict women shrimp gher workers, including skin diseases and fungal and reproductive tract infections, such as leukorrhea, a kind of pelvic inflammatory disease, which causes discomfort and vaginal discharge. His description aligned with the symptoms women shrimp gher workers reported to me, which they described as vaginal infections, "tumors," and excessive bleeding ("*prochur porimane*," one woman said, meaning "enormous amounts"). However, among the women I spoke with in Kolanihat about these medical concerns, none had sought treatment at the hospital, citing the high costs of medical care, which was out of reach for female shrimp gher workers.[18]

FIGURE 5.7. Pradox sign in Paikgachha hospital.

Photo by the author.

Although it is rarely seen in Kolanihat, another source of work for contingent laborers in the shrimp-producing areas is in the collection of wild shrimp fry from the rivers, which is done with very fine blue nets fastened to wooden frames that collectors pull through the water along the riverbank. They then dump the debris that has collected in the net in a metal basin and use a spoon to sort through the twigs and other tiny plant and animal life that accumulated in order to pick out the shrimp fry, which are approximately a centimeter long. They are tiny enough that I myself find it difficult to distinguish shrimp fry from the other varieties of larvae collected in the nets that are visible to the naked eye. This work is extremely low paid and is considered very low status, so it is often carried out by women and children. Quddus reports that 42 percent of female fry collectors are widowed or divorced or else have been deserted by their husbands (making their economic condition particularly precarious) (Quddus 2006). Along with shrimp fry, these fine blue nets catch an estimated 100 to 150 additional species of larval fish and crustaceans; once the debris has been sorted, an estimated 99 percent of the catch is discarded (Ministry of Water Resources, Bangladesh 2001). Because of this indiscriminate collection and disposal of aquatic life, the collection of wild shrimp fry has been blamed for a significant decline in aquatic biodiversity throughout the region (Pokrant 2014) and is likely the cause of much of the disappearance of wild fish stocks. Though the government has banned the collection of wild shrimp fry for these environmental reasons, the ban is largely unenforced.

While women are responsible for *sheola kaj*, work in the ghers for men primarily involves *mathir kaj* ("mud work"). This entails rebuilding the narrow mud embankments demarcating the ghers. These embankments are barely wide enough for a person to walk across with one foot in front of the other, and they rise no more than a foot out of the water. Composed of the silty alluvial clay that covers the land all over the delta, these squat embankments are ephemeral, dissolving haphazardly under the heavy monsoon rains. After the rain, and occasionally in the dry season for more regular maintenance, men are enlisted for a short period of time to repair them by hand, digging into the silky fallen mud, mounding it back on top, and smoothing the ridges with their hands as they go. Their labor is cheaper than a machine, which could likely accomplish the task more quickly and make the embankments more durable (heavy machinery is more commonly used for this work in areas with larger-scale industrial operations than those in Kolanihat). The work itself is needed only occasionally and, like the embankments it produces, is entirely contingent. The earning opportunities it affords are dependent on the vagaries of weather. It does not follow any kind of schedule, unlike agriculture, with its growing seasons, and predictable cultivation calendar, which requires tilling, planting, weeding, and harvesting. This contingency produces deep insecurity, and laborers are unable to rely on it for regular income. It is thus only those laborers whose lives have become most precarious who engage in this work to supplement their livelihoods.

Migration

As suggested in Arjav's story, this dispossession of the rural poor through the expansion and intensification of shrimp culture has resulted in a mass migration from villages like Kolanihat to urban areas. In some communities in Khulna, landless residents survive through seasonal migration for agricultural labor of some male family members, but this is less common in Polder 23, given the lack of subsistence and day labor opportunities to sustain their families in situ (Paprocki and Cons 2014). I heard a handful of reports of people migrating seasonally to the nearby district of Gopalganj, where vast tracts of lowland create the conditions for abundant rice production and high demands for labor. However, most people from Polder 23 migrate to Kolkata, in the neighboring Indian state of West Bengal. Residents explained that at the beginning of the shrimp boom, the majority of these migrants were the newly un- and underemployed landless people. As fewer of these people remain in the village, the flow of their out-migration has waned. Today, it is primarily the young sons of smaller shrimp gher operators who are leaving Kolanihat in search of work. When I asked their families in Paikgachha why their sons migrated, the most common response was "because of poverty." One mother explained to me soberly, "We don't have any work here. We have to eat. Because of that, they are going to India." Another person told me, "They leave as their livelihoods depend on it."

It is worth noting here that in addition to a popular discourse about climate migration from Khulna, there is also a long-standing discourse about Hindu migrants leaving this region due to communal violence perpetrated by Bengali Muslims. While it is certainly the case that religious tensions exist in this region, that communal violence is a problem in Bangladesh in general, and that these dynamics have often driven migration from Bangladesh to India (Alexander, Chatterji, and Jalais 2016; Samaddar 1999), none of the migrants I spoke with cited these as a reason for their migration and it did not appear as a common concern in Kolanihat. While Kolanihat is a majority Hindu village, I regularly heard stories about Hindus as well as Muslims migrating from this area to West Bengal. When I asked residents in and migrants from Kolanihat why people chose to migrate to Kolkata instead of Dhaka, they cited a range of factors including the availability and quality of work, physical proximity, and existing networks that facilitate settlement.

The primary destination for these migrants is a small enclave within the aptly named New Town, a planned satellite city on the outskirts of Kolkata. There could perhaps be no better archetype of "New India" than New Town. The city has sprung up rapidly in the wake of Kolkata's burgeoning IT sector, which claims an annual growth rate of 70 percent (World Bank 2014b, 216). The enthusiasm surrounding this tech boom is manifested by names displayed on apartment blocks in New Town such as "Website Housing" and "TechnoNest." Both the Congress and BJP governments have undertaken programs labeling New Town a "Solar City" and "Smart Green City," representing its important role as a model

FIGURE 5.8. Buildings under construction in New Town.
Photo by the author.

for visions of green urban development in New India (Bagchi 2014; Chakraborti 2013, 2014). The city was formally established in 2007, after the annexation of a vast tract of farmland near Kolkata's airport. In preparation for building New Town, the government of West Bengal used colonial-era land laws to requisition over 7,000 acres of land, displacing 131,000 rural residents (Dey, Samaddar, and Sen 2013, 3). Today it is a patchy urban jungle of sparsely inhabited concrete malls and apartment blocks towering as much as twenty stories high (see figure 5.8). These construction and residential zones are interspersed with residual farmland, which is still being used to graze cattle and for the cultivation of crops in a handful of isolated plots. Besides this, there are few traces of the area's agrarian past, although a small number of signboards posted on roadside shacks proclaiming "Land Losers' Cooperative" are evidence of the dispossession that took place here. These were set up by associations of farmers whose land was grabbed to make way for this urban development. New Town is as much constitutive of visions for India's developed future as Khulna's shrimp landscape is of visions of the same for Bangladesh. Given the relationship between these dynamics of rural dispossession and urban growth, it is fitting that Khulna's migrants should come here in search of new earning opportunities.

Everyone in Kolanihat knows people who have come to New Town, and many of them keep in touch across the border by cell phone. It is through one such contact in Kolanihat that I got in touch with Shonjoy, a young man who came to Kolkata from Kolanihat about three years prior in search of work. His story is typical of those shared with me in approximately twenty interviews I conducted with migrants in New Town.[19] When I spent time with Shonjoy's family at their home in Polder 23, they told me about their struggles with shrimp cultivation that led to his migration. Before the shrimp boom, Shonjoy's father had farmed rice on a sizable (4-acre) plot that they owned, supported by hired labor from four landless neighbors. Debjit, the land grabber from a neighboring village mentioned earlier in this chapter, took control of their land using a similar legal maneuver through which their land title was fraudulently signed away. During the period following the seizure of their land, Shonjoy's father engaged in mathir kaj and other kinds of manual labor that proved to be inadequate for sustaining the family.

Working with neighboring smallholders whose fields had been seized by Debjit, Shonjoy's family attempted to wrest control of their land for about twelve years. Through this collective pressure, the land was finally returned to them. Shonjoy believes that now about 40 percent of the land in their area is controlled by smallholders like his family, usually on small plots of around 3 acres. Before they got their land back, Debjit had absorbed it into his own neighboring gher, where he had cultivated shrimp for the entire period without flushing the salt water during the monsoon. By the time they got their 4 acres back, Shonjoy's family was unsure whether they could do anything but continue cultivating shrimp on it and, moreover, they said they did not have the choice to return to rice farming. Shonjoy told me that his father attempted to farm *aman* rice again about six years after getting their land back. However, his efforts were undermined when vandals (presumably hired by Debjit) cut the small embankments bordering his field in the middle of the night, inundating his rice paddy with saltwater again. As a result, the entire rice crop died. Therefore, he explained to me, despite their great aversion toward shrimp, they felt that rice would not be an option for them under the present conditions. Debjit continued to manage a 130–160 acre gher immediately adjacent to theirs, so there was no way for them to make the decision on their own not to cultivate shrimp, despite having repossessed their land. Shonjoy said that perhaps if they and all their neighbors decided to collectively cultivate rice, they might have a chance of succeeding (though they would still run the risk of having their embankments cut and their fields flooded).

Since then, Shonjoy's family has operated their own small gher on this land. However, they have struggled to make the investments needed to run a profitable gher business. Shonjoy's father took some loans to try to make bigger investments with the hope of greater success, such as buying improved feeds and fertilizers and obtaining superior shrimp fry from hatcheries. Yet despite their efforts, the family repeatedly met with losses. The shrimp business did not provide a sufficient livelihood to sustain the family, Shonjoy explained to me. He speculated that he could have tried to stay in the village and work as a middleman, buying shrimp from neighbors and selling it to traders in Paikgachha town. He said, however, that he did not want to do this kind of work; it is economically risky and vulnerable to frequent attacks by the shrimp virus, and it rarely provides sufficient income to meet a family's financial needs. Faced with a failing shrimp enterprise at home and the family's significant debt to be repaid, Shonjoy left for New Town to earn money to send back to the village. "How else would we eat?" his father asked me, clearly distressed but resigned. This was a common refrain throughout the area, particularly for those whose relatives had been forced to leave to support their families.

When I got ahold of Shonjoy on his cell phone in Kolkata, he told me to meet him one evening at a large, central bus stand in New Town. He seemed cautious but amenable to telling his story, and was excited to hear that I had visited his home and spent time with his family. Shonjoy first came to Kolkata with his older brother, who has since gone back to their village, so now Shonjoy lives there alone for most of the year. When they first came, Shonjoy found work in a power plant, and now he mostly works in construction (see figure 5.9). In New Town, even as a contingent day laborer, he found that he could secure daily or weekly contracts paying as much as 400 Indian rupees (US$6) per day for seven- or eight-hour days. Construction work in New Town's concrete jungle is readily available, and there is a local labor market where contractors explicitly hire Bangladeshi migrant workers.

I asked Shonjoy whether he had thought of going to Dhaka to find work instead of migrating to Kolkata. He said he knew some people from Paikgachha who work in garment factories in Dhaka and he had considered this. Once he went to Dhaka to interview for a job in a garment factory, but he did not get the job and he was unhappy about the anticipated working conditions, which involved twelve-hour days for 8,000 taka ($94) per month.[20] Most of the jobs he could get in Dhaka were garment factory jobs, he said, and they do not pay well. He also placed great value on the work schedule in Kolkata, with most days starting at 10:00 a.m. and ending at 5 or 6:00 p.m. (this schedule is clearly not universal, however, as several other migrants told me they worked until 8:00 or 9:00 p.m.). Yet, there are higher risks and costs involved in traveling to Kolkata than to Dhaka, in particular the 1,600 taka ($19) fee Shonjoy must pay someone to be transported over the boarder across a river by boat. For this reason, he does not return home for visits, although he said that some people do go home for Durga Puja, the most important religious holiday for Bengali Hindus.

FIGURE 5.9. Construction workers in New Town.
Photo by the author.

Even as they enjoy what they regard as a higher standard of living than what they might find in Dhaka, migrants like Shonjoy do not live in the "Website Housing" nor "TechnoNests" that they are building, but rather in densely packed shantytowns on their outskirts. In these settlements, which they refer to as *gram* ("village"), connoting their simplicity, migrants crowd together into small corrugated metal huts and creaky, swaying bamboo structures, many of which are perched precariously over open sewage canals. One such shanty where Shonjoy took me to meet with a group of migrants from Paikgachha was lit by a single light bulb hung from the ceiling and had space for little more than a double bed and three plastic chairs brought in from neighboring dwellings. The air was stifling despite a small electric fan, and a large stack of bedding folded up in a pile on the bed made it clear that at night, the entire space must have been covered with sleeping bodies.

Despite the anti-Bangladeshi xenophobia that infuses the national political discourse in India, Shonjoy explained that the local police do not bother them because they recognized the urgent demand for their labor in the construction of New Town. I asked him if he felt that he was competing for work with Indian labor migrants, such as Biharis, who are considered the other most visible migrant group in Kolkata. He said that there were jobs open to Biharis that he cannot do as a Bangladeshi, such as taxi driving and domestic work in private homes, both of which might require him to present immigration documentation. He said that the demand for the kind of contingent and low-paid labor provided by Bangladeshi migrants was high enough that it outpaced the supply of domestic migrants. Indeed, Mitra's study of migrant workers in Kolkata confirms a hierarchy of informal labor in the city, in which construction work is largely performed by the most contingent, temporary migrants (Mitra 2016).

Shonjoy regards questions about his future pensively. He is not optimistic about Kolanihat's return to rice farming. He said that his father usually comes in November and stays for one to two months; this is the slowest season in the ghers because the cold weather prevents the shrimp from growing, and as a result there is no work to be done then. During this time his father stays with him and they both find work in the daily labor market. Even as his father returns home to continue the gher business, Shonjoy believes that this will not be viable in the long term. Unless somehow the entire village finds a way to begin farming rice again, he explains, there will be no long-term future for them there. The migrants in New Town express mixed optimism about this possibility of return. One young man whose father and brothers have stayed in Khulna to manage a gher volunteered to me that he would return to the village if they could farm rice again. But then he added that he did not think returning would be possible.

Shonjoy's story reveals the economic dynamics driving migration out of Khulna's shrimp-producing region. While these migrations are clearly linked with changes in the local ecology, calling them climate migration obscures the production dynamics that motivate dispossession across a range of rural classes. Thus, understanding the agrarian political economy of shrimp production sheds light on the broader demographic shifts taking place in this region and calls into question claims that massive levels of out-migration from the coastal region are being driven by climate change.

Shonjoy represents the new face of agrarian change and dispossession in Khulna. Through the incursion of shrimp cultivation, even the livelihoods of those families whose landholdings once ensured their economic security have become tenuous. With these uncertain economic prospects, the younger generation is leading an urban transition driven not by personal choice, but rather by economic compulsion. The failure of most development agencies to acknowledge these social and ecological contradictions and the political economies in which they are embedded has facilitated the expansion of shrimp aquaculture. Thus, the antithesis between shrimp and rice cultivation has both material and epistemological dimensions and manifests both socially and ecologically.

While the analysis of this process among NGOs promoting shrimp aquaculture and that of local communities is quite different, the empirics of the transition are, in fact, mostly uncontested. Many development practitioner-researchers, who are aware of the labor transitions taking place in the shift from rice to shrimp, celebrate the increased "productivity" associated with the labor market changes. At a workshop I attended held by WorldFish (one of the primary development institutions driving the expansion of aquaculture in Bangladesh as well as globally) on the topic of "resilience," one researcher described the decline in rural labor opportunities through aquaculture intensification as more "sustainable" because "less people have hard livelihoods" when they are removed from agrarian production systems. This comment is not unique among researchers and practitioners working in aquaculture development in Bangladesh. Even as development agencies express concern for and intention to serve the poorest, aquaculture development programs consistently marginalize people who are landless, which is indicative of a pervasive disregard of these groups as essential constituents of agrarian economies.

This is reflective of a broader vision of rural development (in Bangladesh and elsewhere). Speaking the language of income diversification, development practitioners celebrate the de-agrarianization of rural labor markets and the movement of all classes into nonfarm livelihoods. The manager of one shrimp development program explained to me that the labor market transition resulting from the expansion of shrimp is a benefit to rural economies because it allows people to leave to seek out higher-paying work, as opposed to agricultural work, which is "the least rewarding financially." When I asked him about the impacts on the rural poor who would be displaced, this practitioner candidly responded, "That doesn't concern me," and he explained that reduced labor demand means an increase in profitability. These comments reflect a fundamental difference in understandings of who is and is not part of an agrarian society and who does and does not have the right to be a farmer. Within these perspectives, only landholders can be farmers. This is at odds with the traditional understanding in Bangladesh that all people who participate in agrarian economies are peasants.[21] The poorest members of these communities have historically claimed rights to participate in rural economies as farmers, a participation that was necessary given their very sizable proportion within the agrarian class structure.

This tension was illuminated in another conversation I had over dinner with a foreign scientist from an elite university who was visiting Bangladesh for the first time, supported by a multimillion-dollar grant from a foreign aid agency, in order to study solutions to problems related to climate change in Bangladesh. The scientist explained to me his research team's plans to work in two sites: first, in a rural community somewhere in the coastal zone, and second, in collaboration with garment manufacturers, to help support the growth of Bangladesh's garment industry. The logic of this pairing is quite clear, as the scientist explained to me: "The history of development is the history of moving people out of agriculture." To this scientist, he and his team of researchers can promote what he refers to as "progress" by supporting development that facilitates the transitions of large populations of Bangladesh's rural workers off the farms and into the garment factories. His comments illustrate a much broader vision of development in Bangladesh, of which the expansion of aquaculture is just one part.

6

"WE HAVE COME THIS FAR—WE CANNOT RETREAT"

Adaptation, Resistance, and Competing Visions of Transformed Futures

Benoy came running as soon as he saw me walking down the path that bordered his field, waving his arms animatedly. All of Tilokpur was buzzing, working feverishly: the harvest was good. Better than it had been in thirty years—good enough to confirm what all the residents had hoped but no one had known for sure: that agriculture was still possible in their village (see figures 6.1 and 6.2). It was late April 2015. From approximately 1985 to the mid- to late 2000s, the rice fields of this village had been taken over for shrimp cultivation.[1] Through the collective and ongoing efforts of the landless social movement groups supported by Nijera Kori, Tilokpur smallholders began farming rice again in 2009. Benoy himself had been doing so since 2012 on a 1/3-acre plot that he owns and 1 acre that he sharecrops. The pace of the work on this day was particularly frenzied due to the accumulating clouds in the sky, which had everyone worried about rains rotting the cut stalks of rice before they could be bundled and hauled from the fields. For nine years prior to this return to rice farming, Benoy had been forced to migrate out of Tilokpur to earn money to support his family. He had traveled to a market on the outskirts of Khulna City, where labor contractors hired him to do road construction work. He did not enjoy this work, he said, not least because the working conditions were dangerous and also because it meant he had to stay away from home for most of the year. Now his harvest was good enough that he could stay in Tilokpur year-round.

Benoy's wife, Titash, is a leader in the local landless collectives. Titash has a serene but commanding presence, and though she is soft-spoken, people listen intently when she talks. I once watched her have a spirited fight with an elderly male

FIGURE 6.1. Rice in Tilokpur.

Photo by the author.

leader about whether men or women have more suffering in their lives. In response to an off-handed remark he made that women do not suffer because they do not have to work, Titash delivered a calm but incisive and withering lecture on the political economy of household labor. The lecture ended on the agreeable note that one way they work toward mitigating these imbalances in her household is by all eating meals together at the same time, an almost unthinkable practice in rural Bangladesh, where female members of the family eat meals after serving male family members as a matter of course. Beyond its significance in her own personal life, household labor is, for Titash, critical to Tilokpur's political and existential struggles. During the years when Benoy left the village for work, Titash stayed home with their children and struggled to find ways to feed the family on what meager earnings her husband could send back to the village.

In rural Khulna, though men are traditionally responsible for cultivating rice and other commodity crops, women cultivate the homestead gardens that feed the family year-round and often are also responsible for rearing livestock. During the time of shrimp cultivation, homestead gardening and livestock rearing had become increasingly impossible because the rising salinity crept into the soil of the homesteads from the surrounding shrimp ghers. But now that Benoy was back

FIGURE 6.2. Loading harvested vegetables onto a rickshaw van in Tilokpur.
Photo by the author.

and the harvests were getting better, they had money to invest in livestock again. Titash had been able to purchase three cows that she was successfully raising for milk for the family to both consume and sell. The cows grazed on the grass that had started to grow again as the soil salinity declined and were also fed the chaff from the rice harvest. Titash's garden was flourishing. For her and other women who had stayed in Tilokpur while men had migrated out, the challenges of social reproduction in the shrimp-producing period were just as great as the challenges of agricultural production in the midst of ecological collapse. Indeed, they could not be understood independently of one another.[2]

Benoy was not the only person who left. During the shrimp-producing period, most residents who had previously been dependent on agricultural day labor, particularly the men among the approximately sixty landless households in the village, were forced to migrate out of the village to find work. As one landless laborer explained in 2013,

> Now people have to leave the area to find whatever work they can wherever they can find it, because for shrimp farming you don't need as many laborers so there aren't many jobs. All the people who have to earn

daily wages are still working outside, myself included, despite the fact that shrimp farming has [been] stopped for two years. This is because it has yet to go back to the way it was before. And there are people who keep trying to open the [sluice] gates [to start cultivating shrimp again].

These landless people who left during the time of shrimp production went to Khulna City, like Benoy, or to Dhaka or India. Given the ecological conditions in Tilokpur linked with their migration, it is not difficult to imagine these people being classified as "climate migrants." Given the role of salinity intrusion and waterlogging in impeding continued rice production, they would certainly be considered climate migrants based on the research methodology outlined in chapter 3, under which any migration motivated by a "climate-related stressor" would be classified as climate migration. And yet the conditions motivating their decisions to migrate were clearly much more complicated. By 2015, all these people, including the landless, had returned and found ample employment opportunities in the gradually restored agricultural fields.

History

Tilokpur lies within Dumuria, a subdistrict of Khulna on the inland edge of Bangladesh's coastal zone, in Polder 29. Due in part to its location further inland, the shrimp aquaculture boom came to this area later than adjacent communities closer to the coast. The brackish waters of the coastal region become less saline further inland, making the inundation of agricultural lands with saline water (to fill the shrimp ghers) less convenient. Dumuria is bordered to the west by the Gangrall River, one of the countless tidal channels at the base of the delta of the Ganges-Padma River system.

Tilokpur is a historically rice-producing area, with farming facilitated by abundant rain in the monsoon season as well as an ample freshwater supply from the dense network of rivers and canals, which supports additional crops during the non-monsoon seasons. The most intensive cultivation takes place within a very large lowland area known as Boro Beel (literally "big marsh"), which is surrounded by settlements and canals on all sides. Depending on the exact location of their fields, farmers in this area report that they historically were able to produce two or three crops per year, including rice, lentils, sesame, watermelon, and a variety of vegetables and leafy greens. During the time when shrimp was produced in this area, the low elevation of Boro Beel facilitated the sustained saline waterlogging that is necessary for shrimp cultivation, while also severely impairing continued agricultural cultivation.

Commercial shrimp production began in Tilokpur much like in Polder 23, through the instigation of outsiders with the support of several elites from the local area. As one resident explained,

> We first had the WAPDA [sluice gate], a few influential people got control of it and started bringing saltwater in to start cultivating shrimp. When they started, it was profitable. Then they had hired thugs who would go around the area and take land by force for expanding the shrimp ghers. After taking over the land, they brought in the saltwater by making cuts in the embankments. After a while people began to notice that crops stopped growing. This is because of the salt in the water, nothing grew after a while. Neither trees nor fish grew.... Some people in the area were willing to give 5 *bighas* [1.7 acres] of their land to them, but they took 10 *bighas* [3.3 acres] by force. This is how they took land by force.

The support for shrimp cultivation among some local residents was driven primarily by wealthy landholders from an adjacent village located on the opposite side of Tilokpur from the river. Among these, the most powerful was a man named Imtiaz Saheb, whose wealth allowed him to wield great control over a variety of local concerns beyond the conversion from rice to shrimp. For example, a group of landless group members told me (with considerable dismay) that Imtiaz was allowed to be present at the interviews of all candidates for teaching positions at the local school, leading them to raise questions about several issues in the school's administration (a subject I will explore further).

In addition to water being intentionally brought in to flood the shrimp ghers, at one point the embankment also collapsed because the gher owners had drilled holes in it to install large PVC pipes to bring in the saltwater. This drilling structurally weakened the embankment, and its subsequent collapse caused the entire area to flood. As one resident explained, conflating the intentional and unintentional inundation, "The water rushes in and destroys our homes, kills trees, animals, everything. The shrimp farms are always under saltwater and nothing can survive this water—even roads and houses collapse in the water." The ecological degradation resulting from shrimp farming as well as the unintentional inundation exhibited itself in Tilokpur in a variety of ways (see figure 6.3). The slow death of the trees was the most visible of these ecological impacts and the one cited most frequently by village residents. As one farmer explained:

> The first year that shrimp farming started, the trees were okay, but they stopped bearing as much fruit. By the next year, not only did the fruits stop growing, but the trees started dying. The leaves first started discol-

FIGURE 6.3. The last remaining shrimp *ghers* in Tilokpur in 2012. Although most of the village had transitioned back to rice already, the impacts of shrimp could still be seen at this point. By 2014, these plots had returned to rice with the rest of the village.

Photo by the author.

oring, then the branches started drooping. Slowly the trees themselves died. Our orchard was completely gone.

Despite the ecological crisis that the area experienced due to shrimp production, many farmers attempted to continue growing *aman* rice during the monsoon season, when shrimp production is less profitable due to heavy monsoon rains decreasing the salinity in the ghers. Rice yields were so low during this time that most farmers reported suffering an annual loss. One day laborer described the conditions during this time:

> But the situation with the crops became like this, not even 2 *maund* [74 kg] of rice would grow on 1 *bigha* [.3 acre] of land.[3] I had no work, [so] I would go to areas outside to find work. Unless you saw it for yourself you can't imagine how much I had to struggle to survive. I can't even describe to you how I suffered. I am trying to give you an idea with just my words how hard I had to work to stay alive.

This testimony reflects the importance of labor opportunities in rice cultivation for landless day laborers. In the context of such low yields, farmers growing rice do not require assistance from day laborers when harvesting. Opportunities for sharecropping (on which many landless and land-poor households had previously depended) disappeared, as the shares of rice for both landowners or sharecroppers themselves dwindled to next to nothing and landowners became increasingly reticent to lease out their land under sharecropping arrangements for rice farming.

Despite the struggles that these farmers endured and the exceptionally low crop yields, their persistence in attempting to drain the ghers of saline water annually before the monsoon helped to flush some salt and chemical residue from the soil each year. This practice protected Tilokpur's soils from some of the worst ecological impacts suffered in other shrimp-producing areas, and it may have provided a significant basis for their ability to transition back to rice.[4] Although some researchers have found that annual flushing of salinity from the soil up to a certain concentration is possible, very little is known about whether there is a point at which a return to agriculture is impossible and if so, how and when that point is reached.[5] As one official at the Department of Agricultural Extension put it to me, "We have very primitive knowledge" about the possibility of transitioning back to agriculture from shrimp. The cases of villages in Dumuria and nearby Dacope (Afroz, Cramb, and Grünbühel 2017) that have returned to rice production from shrimp production offer a counterargument to claims that agriculture is no longer viable in Khulna's shrimp-growing regions. I frequently raised these cases with development and adaptation practitioners in response to such claims; they would usually accede the possibility but question whether most communities would have the stamina to survive several years (no one could know how many) of low yields while they waited for their soil to regain its former fertility. However, my interviews with agricultural development practitioners indicate a belief that intrusion of salinity into the underground aquifer would make the return much more difficult, if not impossible, as salinity in the groundwater would for the most part preclude agriculture during the dry season. This would suggest that Tilokpur's aquifer successfully withstood the salinity of the shrimp-farming period and that at least in part, the villagers have this to thank for their successful return to rice.

One of the most revealing conversations that I had in Tilokpur about the discrepancies between rice and shrimp production was with Khokhon, a rickshaw-van driver whom I met one day while he was transporting several bushels of vegetables to the market.[6] As we walked and chatted, I asked Khokhon what he thought about rice and shrimp. I anticipated that as a rickshaw driver, Khokhon's attitudes about shrimp might have been more moderate than those of his neighbors who depended on agricultural labor. Rickshaw-van pulling is often cited as a major form of work for which demand increases within the village in the transition from rice to shrimp because unlike rice, shrimp is harvested year-round and requires prompt refrigeration, so there is an increased need for people to regularly transport what is produced in the village to nearby markets for sale. Indeed, Khokhon confirmed that, as a small but critical link in the value chain for exported shrimp, he made more money during the time shrimp was produced in Tilokpur than he ever had before or after. As a participant in the agriculture value chain, his labor is only required infrequently for transporting the harvest to market, primarily for periods of a few weeks three times per year, and besides that for various ad hoc needs of fellow villagers.[7] And yet Khokhon looked at me incredulously when I asked him this question. During the shrimp time, he told me, everyone in the village experienced *obhab* (scarcity or deficiency). He described *obhab* as a generalized condition that affected the community as a whole, including even the few like him who profited financially. During that time, the only way he could get food was by buying it in the market from people who brought it in from outside, as nothing was growing in the village's fields or gardens. This meant that food prices were much higher. But now food in the market is cheap, he explained, as people in the village are growing a surplus, and he can thus purchase rice and vegetables in the market quite inexpensively or barter the services of his rickshaw. Though the cost of food and other necessities was not out of his reach during the period of shrimp production owing to his increased income from his rickshaw van, it was nevertheless unfavorable for him just as it was for his neighbors. In explaining this shift to me, what he most seemed to want me to understand was that what was good for his community was what was good for him.

Movement against Shrimp

Responding to this environmental degradation in the midst of displacement and personal economic difficulties, residents of Tilokpur faced a protracted struggle as they attempted to return to rice cultivation. One woman told me that they were inspired to act by stories they heard from residents of Polder 23 about the extreme degradation they had experienced. The opposition to shrimp cultivation was spread across a broad base of local class interests. In addition to the landless collectives supported by Nijera Kori, there was also a group that called itself the Saline Water Resistance Committee, which was composed of people who were too wealthy to be eligible for membership in the landless collectives, such as teachers and smallholder farmers. The groups worked together to mobilize a united front against shrimp. Describing their turn toward mobilization in response to collective observations of ecological collapse, one of the participating smallholders described their movement as follows:

> After [the trees died], a few of us got together and decided, with the help of Nijera Kori's landless organization, to close off the gates bringing in the saltwater. We keep closing it, the shrimp farmers keep opening the gate to bring in saltwater. It has been three to four years, and we had to go back and forth closing and opening the saltwater gates. But, finally, we have been somewhat successful. The salt is slowly going away, we have started planting trees, [and] rice is growing as well.

As this smallholder farmer notes, by 2013, the ecological conditions had improved significantly, and Tilokpur had begun to look quite different from the reports of the environmental decline during the shrimp period. While some trees had indeed died, farmers were excited to find that most had begun sprouting leaves again and many were even bearing fruits. Whereas the saline water had killed most of the grass and other plant life in the areas surrounding the ghers, by 2013 most of the landscape was again quite green and lush. The return of cultivated rice as well as grass and shrubs made fodder available, thus creating opportunities for farmers to once again raise poultry and livestock. In 2013, one farmer explained, "Some people have started keeping cows and hens, now that grass has started to grow, but only a few people have started." Thus, Benoy's enthusiasm two years later, in 2015, reflects not only the joy of the successful harvest, but also the success of the protracted struggle that these families undertook individually and collectively, despite uncertainty about the potential for such a robust ecological recovery.

Yet, as the smallholder quoted in the previous paragraph points out, even as they have been successful in their return to agriculture, the movement against shrimp is ongoing (see figure 6.4). In one village next to Tilokpur, where many local residents had also joined the antishrimp movement, farmers woke one morning in April 2015 to find that in the middle of the night, someone had drilled a pipe into the embankment and flooded their land with saltwater to try to force a return to shrimp. Though the residents plugged the hole as soon as they discovered it and the water dried up quickly, the salt, still visible in chalky deposits covering the soil, would make it impossible for them to grow rice or vegetables that season. As a result, men from this village continued to migrate out to Gopalganj, Khulna, and Kolkata to find work.

One small farmer was singled out for harassment and intimidation by a wealthy businessman who wanted to appropriate a piece of his land to build a canal to bring saltwater for a gher in another adjacent village. One afternoon, he showed me a stack of legal documents related to a series of false criminal cases that this businessman had filed against him with the local police. This kind of judicial harassment is a

FIGURE 6.4. Poster for a rally against shrimp on a house in Tilokpur.
Photo by the author.

common means of coercion against the poorest, for whom even the obligation to travel to government offices in a nearby town to contest a false charge can be a punishing burden of time and travel costs. Nijera Kori supports people who are subject to this kind of harassment both through contributing funds to cover expenses related to travel and legal fees and also by connecting members with legal aid, particularly from the Bangladesh Environmental Lawyers Association (BELA), an advocacy group that has led several major cases against shrimp aquaculture at both the local and national levels. This farmer received special support from the local landless groups because their opposition to using the canal for saltwater was linked with advocacy for having it excavated for rainwater collection to expand the irrigation area for rice. They referred to this as the Khal Andolon, or Canal Movement (such struggles over canal excavation are explored in greater detail later in the chapter).

Mobilization beyond Shrimp

While the movement in Tilokpur was catalyzed around a demand to transition away from shrimp aquaculture and back to rice production, its demands did not end there. Understanding how the movement's concerns with ecological destruction caused by shrimp cultivation are entangled with a host of other social and economic struggles is essential to understanding their motivations and, more broadly, the nature of human-environment interaction in the region. This focus draws on a tradition in political ecology that investigates material foundations of environmentalism in peasant and indigenous social movements (Baviskar 1995; Martinez-Alier 2002; Peet and Watts 2004; Rangan 2000; Wolford and Keene 2015). The "environmentalism of the poor," Martinez-Alier explains, derives from a popular ethics demanding social justice among humans, and not from some inherent reverence for nature (Martinez-Alier 2002). Similarly, the movement in Tilokpur has challenged shrimp cultivation on the grounds that it engenders inequitable access to resources and livelihood opportunities among community members, impacts that correlate (but are not synonymous) with its ecological implications. I bring these perspectives on environmental social movements into conversation with those on resistance in Tilokpur in order to lay the foundation for a discussion of alternative perspectives on the possibilities of climate justice in Bangladesh and elsewhere.

In Tilokpur, the landless collectives united to mobilize against shrimp cultivation in their village. They developed strategies and skills for activism that made their mobilization more effective. Sometimes this involved protest marches and demonstrations in regional market centers. They also organized "watch committees" to monitor sluice gates and protective dikes, particularly at night, to ensure that saline water would not be illegally brought in by people hired by would-be shrimp farmers. This often entailed direct physical confrontation by groups and individuals. In addition to these modes of direct agitation, they also learned to use legal channels to support their advocacy. One such channel was the use of the Right to Information (RTI) Act, which was enacted in 2009. The RTI Act gives Bangladeshi citizens the right to request information of public interest from any government or public authority (including NGOs). Several advocacy organizations have begun training citizen groups to use it to hold government agencies accountable for fulfilling obligations to citizens, such as the provision of social security entitlements. Nijera Kori landless collectives have had success with using the law to press officials for information about entitlements to common land access, government decisions about water management, and other concerns that have a concrete bearing on the cultivation of shrimp or rice.

Having learned how to utilize the RTI Act from Nijera Kori organizers, the landless collectives in Tilokpur began putting the principles of the right to information to work in a variety of contexts. The awareness that information was not only their right but was also often publicly available, and moreover that it could be used to demand accountability, became a critical plank in their organizing efforts. For example, the landless collectives suspected that the local schoolmaster was charging excessive fees for administering exams (which prevented the children of landless families from being able to take them), so in November 2014 they enlisted the computer teacher to go online to the school district website and find the table of exam fees mandated by the government. After having this chart printed, they enlisted a Nijera Kori staff member to write a formal letter of complaint on their behalf, explaining that the fees being charged were more than double the amount mandated by the school district and demanding that they be reduced. Following a contentious meeting with an agitated headmaster, who clearly was not prepared for such a sudden shift in power, the fees were adjusted to the government-sanctioned rate.[8] In detailing this protest strategy to me, the members of the landless collectives explained that while they had learned about the utility of securing information for pursuing environmental concerns and securing land rights, these strategies were important in relation to shifting power dynamics concerning a whole range of social and economic issues. Environmental activism to them was only one piece of a broader vision for social justice. Restoration of the environment was also, then, only part of a necessary struggle to pursuing their vision of a more equitable agrarian future. Insofar as environmental activism provided the opportunity for expanding their "repertoire of resistance" (McAdam, Tarrow, and Tilly 2001; Wolford 2010), it provided the foundation for a larger struggle for social change.

Polder 22

The first time I visited Polder 22 was in August 2012, at the height of the monsoon. It poured hard without stopping all day, and I was soaked, my cotton *salwar kameez* and long *orna* hanging heavy and cold. Rezanur Rahman Rose, a Nijera Kori organizer, and I had chartered a boat to take us there, as the roads in this part of Khulna frequently begin to wash away during the monsoon, often becoming completely impassable. For hours all we saw were tall, gray mud embankments fading into the river as we passed island after island. These embankments are formed of alluvial soil rendered barren by the salt that has seeped into them from the shrimp ghers they contain. When we came around a bend in the river and got our first sight of Polder 22, it was, for me, a revelation. In dramatic contrast to the surrounding islands, the embankments of Polder 22 are cloaked in green grass, with trees and settlements rising out of them. There were people too, despite the steady rain, signs of life that were otherwise sparse in the area. This anomalous ecology is the result of decades of struggle by inhabitants of the villages of Polder 22. In the 1980s, when commercial shrimp aquaculture started spreading throughout Khulna through the conversion of agricultural lands to shrimp ghers, in Polder 22 the landless social movement, supported by Nijera Kori, actively opposed it. In 1980, the Dutch development agency in Bangladesh launched the Delta Development Project (DDP) in Polder 22 as part of its overall portfolio of water management programming (Netherlands Ministry of Foreign Affairs 1996; Sur 2010). One aspect of this work involved the construction of a mud embankment surrounding the outside of the large polder embankment. Similar to the traditional *oshto masher badh* (described in chapter 5), this embankment is impermanent; it provides additional protection from storm surges and tidal flooding while also shielding the internal embankment from erosion and breaches. The area between the two embankments was subsequently managed like *khas* land, though it was technically under the ownership of the Water Development Board. Nijera Kori was engaged to facilitate this project, reflecting broader support among some donor agencies in the aftermath of the 1971 Liberation War for more progressive development pathways (in Nijera Kori's case, social mobilization and "conscientization").

Nijera Kori used their role in the DDP to organize landless collectives and help them successfully lobby the government for access to common lands for use in cooperative agricultural production. In exchange for their assumption of responsibility for the manual labor required for maintaining the external embankment, the Water Development Board gave these landless collectives year-round access rights to the ring of land between the two embankments surrounding the polder. Anjan Datta, who worked in Polder 22 as a community organizer for Nijera Kori under the DDP in the 1980s and 1990s (later becoming a development economist), explained to me that the project facilitated an opening for land access that the collectives were able to transform into an ongoing entitlement. "Once they had established a precedent for working on the land," he told me, "they established their rights to it, and showed that they could manage it effectively." This combination of leverage from the DDP intervention with community organizing by Nijera Kori (and activists such as Datta) thus facilitated a transformation in land rights in Polder 22 that has continued into the present.

Collectively, these groups farm indigenous *aman* varieties during the monsoon season. During the dry season they cultivate fish, which they bring in by breaking the embankments at high tide to allow native species to flow in and then trapping them by re-damming the embankments before the tide goes down. These cultivation patterns follow those that had been practiced in most of this region since it was settled until the time when the polders were built (Afroz, Cramb, and Grünbühel 2016; Datta 1998). The common land between these two rings of embankments is thus essential both for the physical protection of the polder and its inhabitants and for the production and social reproduction of many of its most vulnerable residents. The groups plan and perform the embankment maintenance in large teams as well as individually or in pairs or smaller groups throughout the year as needed. When riding in a boat around the polder, it is common to see at least one or two people out with small spades conducting this kind of repair work.

These landless collectives were thus already quite strong and well organized when land in surrounding polders began being rapidly converted to shrimp ghers through widespread, unchecked land grabbing, primarily by elite urban outsiders (Adnan 1991, 1993; Netherlands Ministry of Foreign Affairs 1996). By November 1990, this antishrimp movement had mobilized in opposition to pressure from Wajed Ali, a businessman from Khulna City who intended to acquire over 2,000 acres of agricultural land in Polder 22 to use for shrimp cultivation (Guhathakurta 2008). On November 7, this conflict came to a head in a confrontation between the landless collectives and a mob of armed mercenaries hired by Wajed, during which Karunamoyee Sardar, a local landless leader, was murdered and her body was abducted (Guhathakurta 2011; Primavera 1997).

Karunamoyee's martyrdom thus catalyzed the movement in Polder 22 and beyond. Citing her last words, which are said to have been, "We have come this far—we cannot retreat," Nijera Kori members continue to celebrate her sacrifice both to the landless collectives of Polder 22 and also, symbolically, to the movement against shrimp more broadly. Her death is commemorated annually on November 7 by thousands of members of the landless movement who gather in Polder 22 from all over Bangladesh for songs, speeches, and *jatra* (street theater) performances (see figures 6.5 and 6.6). Today, peasant activists are thankful for the sacrifices of Karunamoyee and other leaders in contributing to the island's ecological resilience in the midst of the surrounding destruction. At the Karunamoyee Day celebration in 2014, a landless group leader from Polder 22 roused the crowd gathered from areas long taken over by shrimp ghers, gesturing toward their fields bursting with the *aman* rice crop and exclaiming, "You can see how beautiful our environment is here in Polder 22. . . . We have goats, cows, chickens, and fresh water. . . . It is because of our leaders that the environment is still so beautiful here!" In articulating this history of ecological protection through collective mobilization, activists enjoin their neighbors from surrounding communities to organize in their own villages against shrimp.

This narrative suggests a radically different interpretation of the drivers of environmental change in the coastal region than that of the popular imagination of climate crisis offered by development practitioners and in media accounts. The activist narrative undermines the "common sense" of climate crisis by suggesting that this crisis is neither natural nor inevitable. More significantly, it emphasizes the role of social mobilization in protecting the environment from degradation. This emphasis also suggests important dimensions of this alternative vision among

FIGURE 6.5. Karunamoyee Day 2014.
Photo by Anders Bjornberg.

FIGURE 6.6. Monument to Karunamoyee Sardar in Polder 22.
Photo by the author.

FIGURE 6.7. Collective rice farming in Polder 22.

Photo by the author.

local activists for the future of this region and for production and social reproduction within it.

Landless groups in Polder 22 continue to pursue this vision of ecological alternatives both through resistance to shrimp cultivation and through a variety of collective agricultural activities. Since the time of the DDP, the landless groups have farmed the land surrounding the embankment together in groups of about twenty (see figure 6.7). Datta explained to me in 2015 that the successful environmental preservation of the island (relative to the surrounding polders) was due to this "strong social organization" supported the project, which persists even today. However, he also said that this powerful work to address the political economy of land tenure through the DDP pilot project was not replicated subsequently by the Dutch government or other development agencies.

In an evaluation of Dutch aid in Bangladesh conducted in 1997, the Netherlands Ministry of Foreign Affairs found that despite the successful organization of landless groups under the DDP, this work was discontinued due to the rapid expansion of shrimp cultivation in surrounding areas and the apparent possibility of it spreading to Polder 22. The report explains:

> In the surrounding polders shrimp cultivation expanded rapidly, putting severe pressure on landowners in the polders to undertake the same activity. Because of this, the evaluation mission questioned the sustainability of the project achievements after its completion. The project was eventually discontinued because the appraisal mission for its third phase shared the same concern, and the Bangladeshi authorities in question did not request continuation. (Netherlands Ministry of Foreign Affairs 1996, 142)

In this way, Dutch development practitioners accounted for an official policy shift from opposing to supporting shrimp aquaculture in Khulna in the 1990s, in the context of significant unrest and contestation over this transition in the region. In January 1993, Bangladeshi political economist Shapan Adnan wrote a policy brief on this decision, which he described as "drastic" and a "U-turn" in Dutch policy in Bangladesh (1993). Adnan, who had been a member of the DDP Appraisal Mission appointed by the Dutch embassy, roundly critiqued the economic analysis cited as the basis of this policy shift, but he also criticized the institutional process through which the decision was implemented. He writes with what he describes as "exasperation" in a footnote to the brief, "I was able to see at first-hand how certain consultants and subject specialists could deliberately ignore evidence (and even flip over from what they confided in private), presumably in order to ensure the renewal of their contracts" (1993, 1). In this note, Adnan's account conveys the frustration described by many activists with the complicity of aid agencies in the promotion of shrimp aquaculture in this moment of agrarian transition. Whereas the transition to shrimp aquaculture today is often treated by development agencies as inevitable, this account indicates how it arose instead through deliberate decisions made in the face of widespread contestation.

Despite this discontinuation of support, the landless groups organized under the DDP have remained strong, in part through the organizing assistance Nijera Kori has continued to provide. For about four months out of the year, they grow *aman* rice between the double ring of embankments surrounding the island. During an additional six months, they raise fish from a diverse mix of indigenous larvae that flow in from the adjacent river. Some of this fish and rice is sold in nearby markets, but much of it is also consumed by the families of the landless group members, among whom the remaining harvest is split up equally. One member told me that her share of the rice yield was usually enough to feed her family for six months (which supplements the rest of the families' independent earnings and production, either from work within or outside the village or what they grow in their own personal small plots or gardens).

Another critical aspect of this alternative vision of the agrarian future of Polder 22 is the commitment of landless groups to cultivating indigenous varieties of seeds for rice and other crops. Since the time of the Green Revolution, one of the key tenets of rural development in Bangladesh (as elsewhere) has been promotion of hybrid seed varieties, which are either imported from outside the country or developed within the country, often using foreign technology (Feldman and McCarthy 1984; Gupta 1998). The principal merit of hybrid seeds is said to be that they result in higher crop yields relative to indigenous varieties. Among NGOs working in climate change adaptation, hybrid seeds are particularly celebrated as a solution to addressing some of the worst climatic vagaries, such as drought, salinity, flooding, unpredictable rains, and uncertain seasonal change (Mackill et al. 2010).[9] Yet many farmers in Polder 22 have not shifted toward these new "climate-smart" hybrids, but instead have embraced traditional varieties and are working collectively to revive them. Whereas farmers have found that crops grown from hybrid seeds can be destroyed in conditions of excessive flooding, several farmers described to me indigenous strains of rice that within a matter of days would naturally grow high enough to reach above flood waters when submerged. They also described traditional varieties that are naturally saline tolerant and thus better adapted to the brackish waters of the coastal flood plains than the new hybrid strains promoted by agricultural extension agents.

Farmers in Polders 22 and 29 cite this natural adaptive capacity of indigenous rice strains as a major reason for their growing interest in cultivating them today. This shift in cultivation has required a collective mobilization to revive varieties that had begun to decline in use and that do not enjoy the support and dissemination of government and NGO agricultural extension agents. This mobilization has primarily taken place through the development of community seed banks for saving and sharing local rice and vegetable varieties (see figure 6.10). Though some organizers described to me visions of more formalized institutional spaces for preserving and propagating the use of these seeds, the approximately half-dozen seeds banks I observed in the area were housed informally in people's homes (see figures 6.8 and 6.9). Rice seeds are often stored in traditional earthen vessels, while less ubiquitous vegetable seeds are kept in repurposed plastic bottles, bowls, and crumpled-up pieces of newsprint and fabric. Through an ad hoc system of revolving dispensation, farmers deposit particular seed varieties after a harvest, borrow from the bank again the following season before planting, and then redeposit a portion of their successfully harvested seeds. The success of this system is contingent on the unique feature of indigenous seed varieties to be saved from season to season, unlike hybrids, which lose the unique characteristics for which they were bred when saved between generations, which compels the farmers to buy new seeds annually.[10]

FIGURE 6.8. Home seed saving in Polder 22.

Photo by Anders Bjornberg.

Canal Movements

Another critical component of the agricultural capacity in both Tilokpur and Polder 22 involves the presence and excavation of canals that, in addition to facilitating drainage, also store and supply freshwater throughout the year. In Polder 22, the capacity at the northern end of the island for farmers to grow winter crops is facilitated by a canal that runs diagonally through the polder. The canal connects with the river through a sluice gate on the west side of the polder that is opened during the monsoon, when it fills with freshwater (both from the river and from the natural drainage of rainwater into it). However, at the southern end of the polder, the canal has filled with sediment, which has caused its bed to rise, its width to narrow, and its capacity to reduce to such an extent that it quickly dries up shortly after the monsoon. This process of sedimentation is an unexpected

FIGURE 6.9. Bottles containing traditional seeds saved by farmers in Polder 22.

Photo by Anders Bjornberg.

FIGURE 6.10. Nijera Kori collective seed bank in Polder 22.

Photo by Anders Bjornberg.

result of the polder system (examined in chapter 4). The reduced velocity of water within the canal (caused by the hydraulic system inside the polder being cut off from the outside) results in a reduced capacity to flush the sediment out, with silt being deposited within the canal instead of being pushed out to the river. As a result, there is a lack of available freshwater at the southern end of the island during the winter, which means that most of this land lies fallow during the dry season.

The remedy for this sedimentation problem is fairly straightforward: the sediment can be excavated from the canal. This can be done either by hiring local laborers (certainly the preferred option among local residents, as it generates additional employment for the poorest) or through the use of dredging machinery. Excavation of the canals is among the greatest demands I have heard from villagers throughout the coastal region, in a variety of different ecologies and production systems. In the parts of the coastal zone where the rivers become saline in the winter, this storage of freshwater can add an extra crop to their annual production calendar, meaning that farmers can as much as double their agricultural incomes over the course of the year. At a conference on the Ganges Coastal Zone organized by the CGIAR consortium in 2014, one researcher from the International Water Management Institute (IWMI, a member of the consortium) cited an unpublished study indicating that these canals can store enough water to irrigate a full crop of winter *boro* rice in the coastal zone and that this area could be increased to 40 percent if the canals were properly dredged. The researcher argued that these findings undermine common perceptions that water resources are a constraint to production in the coastal zone. At this same conference, a researcher from the Bangladesh Agricultural Research Institute (BARI) explained that even canals filled with water of low to medium salinity could be a good source of irrigation for more saline-tolerant winter crops such as wheat, mustard, and watermelon. These findings shared by Bangladeshi researchers are significant in that they demonstrate possibilities for supporting agricultural expansion in the coastal region through local labor and low-tech physical interventions. They also again undermine claims that agriculture is no longer viable in this region.

The major challenge to pursuing these options is a vacuum of responsibility for carrying out this essential infrastructural maintenance. This represents a long-term entrenched failure resulting from the ambiguous division of responsibility between donors and the government in short- and long-term development intervention; we already observed this dynamic in the inception of the polder system in the 1960s (described in chapter 1). The Bangladesh Water Development Board (BWDB) is responsible for infrastructural construction but the scope of their responsibility for the maintenance of exceptionally large infrastructures is less clear. This maintenance could also be considered the purview of the Local

Government Engineering Division (LGED), although this agency is also responsible for construction and its focus on maintenance is usually confined to larger types of infrastructure such as roads, cyclone shelters, and certain large embankments. Some among the donor and government community say that local communities should be responsible for dredging their own canals. The responsibility vacuum is compounded by the fact that donors are sometimes willing to fund maintenance as part of their limited-term project activities in a given area, so the possibility of this additional funding discourages government commitments to maintenance. Despite sometimes funding such activities, donors consistently deny responsibility for them, instead expressing a preference for new and novel interventions and short-term commitments. When I talked to an official at a large USAID program for climate change adaptation in Khulna, he explained in detail how community members played a key role in planning the project interventions, including the kinds of work that would be carried out under the auspices of the project and what production systems would be promoted as a result. However, when I asked about what the boundaries might be of potential work to be carried out under the auspices of the project, he explained that water management (including canal dredging) would not fall under the possible scope of supported interventions. He described one community that had requested that the available project funds be spent on excavating a reservoir (to provide water for irrigation and domestic use, like a canal). However, the project staff believed that such an intervention would not improve villagers' incomes (meaning income beyond agriculture), and therefore they instead decided to connect the community with a commercial aquaculture project instead. Here we see a clear conflict between demands of local communities for basic maintenance to support agrarian livelihoods and the visions of a development agency for transitioning toward nonagricultural livelihoods, with a preference for experimental new technologies. These are hallmarks of the adaptation regime in Bangladesh.

In 2014, I attended a Consultation Workshop hosted by the FAO in Khulna to solicit feedback on a new project to be funded by the Global Environment Facility (GEF) to promote aquaculture as a strategy for adapting to climate change.[11] Held in a conference room in Khulna City's nicest business hotel, the meeting was attended by about twenty invited local government bureaucrats and some faculty from Khulna University, along with a group of Bangladeshi and foreign consultants hired to plan the project. One consultant explained to the gathered local experts, "We're most interested in finding out what technologies are most appropriate, what kinds of protocols to recommend, like [shrimp-]stocking densities [in ghers]." Despite the consultant's request for purely technical recommendations about interventions that had already been planned, several of the attendees wanted to talk about the conflicts between rice farming and shrimp production and related needs for alternative water management systems. Several mentioned the need for canal dredging to improve freshwater storage for irrigation. An official from the Department of Fisheries insisted to the consultants that "canal excavation should *definitely* be a component" of the GEF project, explaining that lack of access to water due to canals being filled with silt was the primary obstacle to production for farmers in Khulna. These recommendations were at odds with the plans to support aquaculture that had already been developed for the project and approved by the GEF, and were therefore not of particular interest to the project team. Another of the consultants weighed in:

> It's good to brainstorm, I have no objections to that. But we need to identify what actually needs to be done. If waterlogging [related to shrimp ghers] is the key issue, then how to identify problems and potential solutions? And then what kinds of problems might be created downstream? If it cannot be solvable, what can be done to get the farmers to adapt to the problems. If the problems are intractable or very costly [to address], then we should help farmers adapt regardless.

This statement recalls the conflicts over waterlogging described in chapter 4, in the sense that it suggests that instead of mitigating waterlogging, they should find ways to make farmers adapt to it, especially where it is caused by shrimp production. It also speaks to the recurrent insistence that agriculture is no longer viable in this area and farmers should learn to accept and adapt to shrimp aquaculture. These widespread discourses about the unviability of agriculture do not take into account the potential for expanded irrigation through improved canal maintenance. As we see in Tilokpur and Polder 22, the excavation and storage of freshwater in these canals extends the possibility of irrigation during the dry season, thus facilitating year-round agricultural production.

Ongoing Mobilization

The testimonies celebrating the successes of these mobilizations against shrimp aquaculture should not be mistaken for lack of local contestation. Indeed, the movement to continue farming rice is ongoing and entails vigilant and persistent organizing efforts. I met two people in Polder 22 who told me they had formerly been members of the landless collectives but left because of their aversion to the continuous agitation required of membership. Though this was not common, it did reflect the tireless work demanded by these struggles. One man told me he had left because he did not like the constant fighting required of movement members, and others later mentioned to me that this man was very "peaceful" and therefore was not suited to the ongoing work of mobilization. One woman told me she had left because there was ongoing violence directed at members from outsiders and some local elites and she was afraid of the conflict, so she thought it was better to stay out of it. Another man, who lived on a marginal piece of land on the embankment and had moved to Polder 22 when he was pushed out of his home village in another polder by shrimp production, told me that he disliked the landless collectives because they controlled the land around the outside of the embankment together, and although he could have joined one of the collectives, he would have preferred to have had a plot of land to himself. Nevertheless, even those who choose not to participate in the ongoing mobilization benefit from the continued agricultural production in the polder, and these expressions of disinterest in the movement did not translate to dissent against the agriculture it promotes.

The ongoing work of mobilization takes many forms beyond the continuous collective cultivation undertaken by landless group members, though today it is often more mundane than spectacular, unlike, for example, the attack in 1990. One example of this was the introduction of a program funded by the Dutch embassy known as "Blue Gold" beginning in 2013. While the program was technically implemented by Bangladesh government agencies, it was planned and managed by a group of consulting agencies led by the Dutch firm EuroConsult Mott MacDonald. Polders 22 and 29 were two of several polders in which Blue Gold had planned interventions to promote new agricultural technologies, market linkages, and improved water management infrastructures in the coastal region.[12] I met several times with staff involved in this program in both Dhaka and Khulna who told me repeatedly that they had no plans to promote shrimp aquaculture in Polder 22, and some articulated their opposition to the inequitable agrarian political economy of shrimp production. Yet I also repeatedly heard doubts about the program from landless group members in Polders 22 and 29, who insisted that they believed the program would in fact support the introduction (or reintroduction) of commercial shrimp cultivation in their villages. Regardless of whether this was the intention of the planners and administrators of the program, I heard reports in both communities that the program had targeted as beneficiaries the local elites who had opposed their movements against shrimp. In supporting these local elites, movement members reasoned, Blue Gold was supporting the continuation of shrimp aquaculture, regardless of the intentions stated in their program documents.

It was clear to me that these assumptions of Blue Gold's interest in promoting shrimp on the part of movement members were based on the evidence of experience and were in this sense logical even if program staff continuously insisted on the contrary. In one particularly troubling incident in Polder 22, a landless movement member told me he had been attacked after dark in a small field he was cultivating by Blue Gold members who wanted to take over the landless groups' collectively cultivated lands and turn them into shrimp ghers. His broken arm, which hung in a sling, reflected the seriousness of this attack. As I spoke to other community members and Nijera Kori organizers about this incident, a modified version of this interpretation of the event seemed more likely. Indeed, there is anger among wealthier community members about the landless groups' near-monopoly of the land around the outside of the polder—although landless people have a constitutional right to *khas* lands, in many communities this right is co-opted by elite residents (Iqbal 2019). Even as these residents may also be members of Blue Gold, their interests in these lands, or even possibly in shrimp cultivation, are not necessarily related to their membership in the program. I did hear from some landless group members that they were also members of Blue Gold as well (although I also heard reports from some landless members in Polders 22 and 29 that they had been explicitly excluded as beneficiaries because of their activism against shrimp).

This ambiguity and confusion over the goals, plans, and beneficiaries of Blue Gold could have been cleared up through direct and concerted communication about the program between all parties. Yet in 2015, in frustration that the information was not forthcoming, landless groups in Polder 22 submitted an RTI Application to the Khulna Blue Gold office requesting details about the work proposed for their community under the project's purview. They received support in filing this application from local Nijera Kori staff, including completion of the application and transportation to the Blue Gold office, where it was delivered. When Blue Gold staff failed to supply the requested information, the landless group members appealed their request all the way up to the national RTI arbitrator, who ordered Blue Gold to provide the relevant documents. When the details were still not forthcoming, the landless groups appealed to this national body again and were finally given some documentation from Blue Gold officials, although they found it to be sparse and not particularly detailed, leading them not to trust its veracity and completeness. In discussing these events with one Nijera Kori organizer, I said that I had personally received more detailed reports from program staff when I visited their offices, which I offered to share with the groups in the hope of clearing up what I saw as confusion over whether Blue Gold would be promoting shrimp. The organizer looked at me as if I had misunderstood the point of the groups' efforts and said, "They have to give the information directly to the landless groups; it is their right to request it and to have it." Indeed, their efforts were directed not only toward knowing and understanding plans for development interventions in their community, but also toward securing the right to active participation and self-determination in creating those plans. This was a point that appeared to be lost on the staff of the Blue Gold program: it did not matter whether they felt they were acting in the interests of all community members; rather, what mattered to these collectives was that they had a role in determining what those interests were and how they would be pursued. Without this equity and transparency in planning development interventions, the program was actively eroding political enfranchisement and self-determination among this community's poorest residents, which they had struggled to secure and maintain over several decades.

These concerns turned out to be very well founded. Around 2015, Blue Gold started to form new constituent groups to manage the land around the polder and to experiment with new cash crops and seeds introduced by the program. These groups included a small number of landless people along with wealthier local residents and some absentee landlords. Subsequently, Nijera Kori organizers reported to me that these groups had supplanted the landless collectives, causing them to lose access to the land they had been farming for decades. Without the strong social organization of the landless groups, which not only cultivated the land but also maintained the embankments, the infrastructure began to deteriorate at an unprecedented rate. These organizers told me that they lobbied the program to repair the damaged embankments but repairs did not come soon enough. In May 2020, Cyclone Amphan caused widespread damage across the coast of eastern India and Bangladesh. Organizers reported to me afterward that when the cyclone hit Polder 22, it caused the degraded embankments to breach and the entire island to flood. All the crops in the fields were lost. During two previous cyclones, Sidr (2007) and Aila (2009), the embankments of Polder 22 had been well maintained enough to withstand such damage (while the embankments of all the neighboring polders had broken). With the Blue Gold program ending in June 2020, the landless groups were able to successfully lobby the government to have the leases of the land returned to them, and they began making plans to repair and maintain the embankments with the hope of returning to collective cultivation.

This experience of the struggle of the landless groups to understand and have a say in the interventions being planned and taking place in their community is an example of how the ongoing agitation against shrimp production transpires in Polder 22. Even as the overt violence of the early period of land grabbing by shrimp producers has largely receded, the maintenance of their enhanced control in the production system requires ongoing vigilance and action. The landless group member who had been attacked explained to me that were it not for the collectives, the landless people would have all disappeared by now, and added, "We're not afraid of the *lathi* [club] because we know our rights."

Incremental versus Transformational Adaptation

The modes of resistance and collective mobilization demonstrated in Polders 22 and 29 reflect a particular imaginary of the future of this region that is at odds with the future imagined by many development agencies and other actors who participate in the adaptation regime. The political economy of rice and vegetable production in Tilokpur, as well as in Polder 22, demonstrates a vision of the possibilities for rural futures imagined by local communities that has been undermined both discursively and ecologically by the enthusiasm for shrimp production among development agencies. In particular, the vision from these villages in Khulna demonstrates a transformation toward agrarian futures, not away from them.

These divergent visions and trajectories of transformation are significant in light of widespread calls among adaptation practitioners and academics for a focus on "transformational adaptation" (Anika Nasra Haque, Dodman, and Hossain 2014; Kates, Travis, and Wilbanks 2012; O'Brien 2012; Pelling 2011; Rickards and Howden 2012;). In framing the scope of the activities carried out under the banner of adaptation, practitioners and policy makers have increasingly begun to refer to the distinct categories of "incremental adaptation" and "transformative adaptation," which are understood to be discrete paradigms addressing different kinds of interventions and different time frames of change. The IPCC Fifth Assessment Report defines the scope of and difference between these paradigms as follows:

> Incremental adaptation refers to actions where the central aim is to maintain the essence and integrity of the existing technological, institutional, governance, and value systems, such as through adjustments to cropping systems via new varieties, changing planting times, or using more efficient irrigation. In contrast, transformational adaptation seeks to change the fundamental attributes of systems in response to actual or expected climate and its effects, often at a scale and ambition greater than incremental activities. It includes changing livelihoods from cropping to livestock or by migrating to take up a livelihood elsewhere, and also changes in our perceptions and paradigms about the nature of climate change, adaptation, and their relationship to other natural and human systems. (Noble et al. 2014, 839)

While incremental adaptation is understood to be the current modus operandi of climate change adaptation, calls for a paradigm shift toward transformational adaptation arise in every discussion of the path forward for adaptation in

Bangladesh as well as globally. These discussions typically involve a distinction between incremental adaptation as a short-term process (often referred to as "addressing symptoms") and transformational adaptation as a long-term process aimed at more fundamental and systemic change, while recognizing that pursuing both strategies simultaneously will be necessary. "Transformation" has become the word used to indicate a change in development priorities, and it often indicates a different vision for the future to be pursued by development agencies.

The notion of transformation carries the implication of a normative progressive orientation in both popular and scholarly usage, and many academics have also explicitly identified the concept's emancipatory possibilities (Bahadur and Tanner 2014; Bassett and Fogelman 2013; Eriksen, Nightingale, and Eakin 2015; O'Brien 2012; Smucker et al. 2015). In this vein, at a conference on climate change adaptation in Kuala Lumpur in 2014, Farah Kabir, the country director of Action-Aid Bangladesh, described transformational adaptation as "the new way to development" and explained that it would entail a "redistribution of power." Yet, like many development buzzwords before it (Rist 2007), transformational adaptation—what it means, and how to pursue it—remains ambiguous and elusive. We must ask, "Transformation to what? For whom, and how?"[13]

Incremental Adaptation

In Bangladesh, programs reflecting the incremental adaptation paradigm often involve technical interventions with roots in a "business as usual" development approach. These projects often involve relatively minor technical interventions, such as the introduction of hybrid rice seeds that are considered to be more saline tolerant or the promotion of homestead container gardens for growing vegetables in areas where the soil has become too saline to plant gardens in the ground. They are, by definition, pursued with the goal of intensifying or sustaining status quo production dynamics and power relations. For example, one report prepared by BRAC for UN Women proposed alternative livelihood options that would be appropriate for promotion as climate change adaptation strategies among women in each of 20 upazilas in the coastal zone (BRAC 2013).[14] In Paikgachha, their high-priority selected interventions involve the introduction of non–climate sensitive livelihoods (in other words, nonfarm livelihoods), including manufacturing of handicrafts (for example dolls, handmade paper, and other decorative household items such as those sold at BRAC's urban department store chain, Aarong), as well as interventions that support the "extension" of aquaculture, including the collection of shrimp fry from rivers and the repair of nets and cages used in aquaculture. These strategies are contingent on and bolster the continuation of the particular relations of production currently existing in the shrimp-producing area.

This is precisely what makes them incremental adaptation strategies. Yet, in the sense that they also entrench dynamics of dispossession from agricultural livelihoods, they might also be seen as transformational.

Perhaps the clearest example of the effects of incremental adaptation can be found in southern Shatkhira, in the area surrounding Munshiganj, to the east of which lie Paikgachha and Dumuria. Shatkhira is iconic for its crises and its failures. After it was devastated by Cyclone Aila in 2009, donors and NGOs flocked to the area to establish new programs for climate change adaptation. Today the remarkable density of these programs is made visible by the signboards that line Satkhira's main highway, all the way to Munshiganj, where it ends in a dusty road through a small market and a handful of tea stalls surrounding a wall of NGO and foreign embassy logos, photographs, and signs pointing in different directions to various adaptation program offices and demonstration sites. The Climate Smart House described in chapter 2 is an example of one of the most celebrated and publicized incremental adaptation interventions in Munshiganj. While there is a great deal of celebration of reaching project targets and implementation goals through programs in this area, one donor confided to me that he considered this region to be a "graveyard of failed projects," a statement reflecting both the sense of failure of many adaptation interventions as well as the great abundance of experiments that have made their way through the area. Moreover, the comment indicates that the adaptation regime, more than producing any particular "successful" strategy for sustainable development and continued habitation of this region, generates precisely this landscape of intervention itself—a never-ending stream of experiments with no future. Incremental adaptation programs are significant for what they represent; in their aggregation, they propose and enforce new ways of governing this landscape.

Adaptation experts with whom I spoke about the time frames of these interventions estimate that the impacts of incremental adaptation will last for somewhere between zero and twenty years (some cite the figure as zero to five years, while others say two decades might be possible), whereas transformational adaptation is designed and anticipated to generate changes that are sustainable in the much longer term. There is no clear consensus on these figures; they can be based on both predictions of future environmental change and normative planning assessments. Some scientists have ventured into the intersection of both, such as one recent study in *Nature Climate Change* that estimates a potential time frame for the necessity of transformative adaptation of twenty to fifty years (Rippke et al. 2016). The estimated time frame of these interventions reflects expectations about the sustainability of rural livelihoods in these areas in general. Incremental adaptations in coastal villages in Khulna are not expected to last longer than two decades because for many adaptation practitioners, these spaces themselves and the communities that inhabit them are often not expected to last longer than that.

Transformational Adaptation

Development practitioners working in Bangladesh to promote climate change adaptation also frequently discuss the imperative for transformational adaptation. Many experts cited to me the need for "social engineering" to effect transformational change, though either deliberately or implicitly, they avoid specific descriptions of how this engineering will be carried out and interventions that it might entail. While calls for a shift toward transformational adaptation abound in Bangladesh, concrete examples of existing transformational programs are rarely given. The need for transformation in socioecological systems that are perceived to be threatened with (or already experiencing) dystopia is considered to be self-evident; however, the nature of the transformations necessary to avert this is far more nebulous.

Yet, when examining descriptions of potential transformation, one recurring theme emerges among these nebulous visions. In Bangladesh, the role of rural out-migration in imagining this vision of the future through transformational adaptation is stark, and rural-urban migration is one transformational adaptation strategy that is consistently cited. In every conversation I had with development practitioners about what transformational adaptation could look like or how it might be pursued, they discussed the critical role of rural out-migration from Khulna. Options cited for interventions to support this migration included "planned migration" and training farmers in skills more appropriate to work in urban areas as well as transforming production from agriculture to shrimp aquaculture and promoting related transformations of rural economies to urban ones. One DFID official explained that the agency was reorganizing its programs in order to specifically address transformational adaptation, and that this would entail skills training for "transformed livelihoods" and migration from the Southwest to the urban areas.

This reliance on migration as an example of potential transformation is indicative of the broader normative dimensions of how transformational adaptation has come to be understood. The idea of transformation serves to arbitrate the kinds of futures that are deemed possible and necessary. Of all the kinds of socioeconomic and ecological transitions that may be possible in southwestern Bangladesh, when it comes to discourses of transformational adaptation to climate change, moving people out of their communities appears to be the only conceivable option cited. Thus, transformational adaptation is indicative of the broader dynamics of how the adaptation regime is governed. While containing multiple and sometimes conflicting perspectives, it also governs the boundaries of the kind of futures or livelihood strategies that are considered viable and worth pursuing.

While the idea of the need for transformation is talked about with a high degree of confidence, when it comes to specifying what this might look like in practice, most adaptation experts become more circumspect. Many people discuss the need for a combined approach involving both incremental and transformational interventions, pointing out the challenges that true transformational change will entail (for recipient communities rather than for practitioners). Many cite the inevitable disruption that will be necessary in rural communities in the Southwest in order to achieve adaptation. One practitioner at an NGO adaptation program explained to me frankly, "You cannot do things overnight. People will suffer. Things will have to change." Discussions of necessary transformations thus entail assumptions not only about their inevitability, but also about the dispossession that will take place through and as a result of them. Another adaptation expert, reflecting on his organization's advocacy for recognizing migration as a strategy for adapting to climate change, explained to me that the need for transformation also entails a "need to reorganize societies." In order to do this, he said, it will be necessary to confront the fact that "there will be winners and losers."

This phrase, "there will be winners and losers," is repeated frequently among climate experts who are invested in establishing transformational adaptation as the new adaptation paradigm. In this way, it has come to be a kind of orthodoxy of the adaptation regime. It is also reflective of the understanding that there are already winners and losers. Indeed, the people living in what is understood to be Khulna's dystopic ecology today are clearly among those who will be the "losers" in this emerging paradigm. However, there is some slippage in the nature of how these losses are understood and discussed. While their supposed inevitability is attributed to the dynamics of climate change itself, the discourse of transformational adaptation indicates that the actual dynamics of dispossession will be carried out through adaptation programs, interventions that are by no means a foregone conclusion.

The visions of agrarian change pursued by these collectives in Polders 22 and 29 offer a radically different vision of the possibility of rural life in Khulna in the time of climate change than those pursued through the adaptation regime. The members of the landless collectives who mobilized against shrimp aquaculture and continue to resist the transition to it have drawn on their lived experiences in these communities to develop alternative visions of transformation for life and livelihoods. In this sense, the "spatial imaginaries" of these communities are constituted precisely through the particularly situated experience of agrarian production itself (Wolford 2004). Discrepancies between the visions of life in a climate-changed future that draw on spatially situated lived experience and those that appeal to a geographically disembedded development teleology thus highlight the key tensions in the adaptation regime. The different spatial imaginaries articulated by adaptation experts and local residents suggest competing moral economies of climate governance.[15]

In reflecting on the adaptation regime more broadly, what is perhaps most conspicuous about the social movements described in this chapter is that they do not articulate their demands as a desire for "climate justice." Indeed, climate change as such does not factor into their movement narratives of social or ecological change either historically or in the future. Yet their efforts to secure more just agrarian futures shape their survival in the time of climate change, and thus fundamentally politicize climate futures. In this sense, they are embedded in the same normative framework that Borras and Franco refer to as "agrarian climate justice"—goals of redistribution, recognition, restitution, regeneration, and resistance (2018). Political demands for climate justice (Forsyth 2014; Ranganathan and Bratman 2019; Routledge 2011, 2016) necessarily intersect with a variety of other localized struggles over power to shape socioecological change (Cohen 2016, 2017; Elliott 2017, 2019; Koslov 2016). "Transformation" does not take place in a vacuum; it transpires within a broad landscape of competing political visions and drivers of socioecological change. In demanding transformations that create just and equitable agrarian futures for their communities, the landless collectives described in this chapter articulate a strong vision of agrarian climate justice.

CONCLUSION
Climate Justice and the Politics of Possibility

> "The apocalyptic scenarios did not help. . . . I found my mind finally beginning to turn from the elegiac *what have we done* to the practical *what can we do?*"
>
> —Zadie Smith, 2014

Let us consider the possibility that the crisis of climate change offers the opportunity for the radical redistribution of power—perhaps even demands it. If climate change exposes the fundamental crises of capitalism and their inherent inequity, then it also exposes the need for transformative change in our political economic systems. Adaptation to climate change is a tool for redistributing power and resources: as we have seen in this book, it produces winners and losers. Who benefits from this redistribution, and how, is mediated by existing systems of power. Yet these power structures are not totalizing. They can be resisted.

Forging new visions of climate justice through which a more equitable redistribution of power may become a possibility requires a more expansive understanding of climate change. It requires denaturalizing the effects of climate change and situating them within their social, historical, and geographic context. It also requires careful attention to existing and emerging political imaginaries that equip us with alternative ways of thinking about possibilities for the future. How can we learn from social movements working toward decolonization and class, race, and gender equality to forge new understandings of what climate justice might look like?

The landless collectives in Khulna whose struggles I examine in chapter 6 point us to precisely such alternative political imaginaries. Where the adaptation regime sees inevitable crisis in their communities, they see political and ecological possibility. Where some adaptation programs have pursued the entrenchment of unequal agrarian political economies dominated by shrimp aquaculture, they have fought for the redistribution of land and resources to make agriculture a possible livelihood for all members of their communities. Through this politics of possibility, they have forged a broader, more inclusive, and more equitable vision of climate justice.[1]

In Bangladesh today, many of those who are invested in the adaptation regime locally and are also active in the climate movement globally have made demands at the international level that draw on dystopic imaginaries of Bangladesh's future climate crisis. They have mobilized these dystopic imaginaries in global climate negotiations to demand both reductions in global carbon emissions and reparations for what has already been done in the form of adaptation funds. Yet, in doing so, they have often allied themselves with more destructive development logics that naturalize crisis and dispossession and undermine local imaginaries of social justice. This strategy is not necessary, and neither is it inherent in the demands for global climate justice. Even if climate change will pose major threats to Bangladesh sometime in the future, that does not mean that the climate justice movement cannot also ally with social movements that are making demands for justice in the present and imagining possible alternatives.

Imagining the future from communities in Khulna contains the conditions of this possibility. The movement against shrimp that is spreading—slowly but surely—throughout Khulna is shaped by fundamentally optimistic political imaginaries that predict a transformed political economy and a revived ecology. For these local activists, resistance to the adaptation regime and climate crisis entails an insistence on the possibility of continued agricultural production in the coastal region. Communities have mobilized both to continue farming rice and to return to rice cultivation in places where shrimp production has already begun. They embed these demands in broader political visions including gender justice and the systemic redistribution of power and resources. By insisting on rice agriculture and the broader dynamics of production and social reproduction that accompany it as a real alternative to shrimp, these communities contest the teleology of climate crisis. In doing so, they contest the dystopic spatial imaginary of the adaptation regime itself. They show us that this imaginary of the adaptation regime is not totalizing. There can be visions of climate change adaptation outside it. The success of social movements in Khulna offers possibilities for such alternative climate imaginaries, helping us to see beyond the ideology of the adaptation regime.

These imaginaries are grounded in local experiences and political struggles, but they are simultaneously globally inflected. The social movement groups in Khulna that struggle against shrimp are aligned with and supported by an international solidarity movement campaigning against shrimp aquaculture in communities around the world. From Latin America to Southeast Asia, this coalition pursues an end to the social and ecological destruction produced by shrimp cultivation, motivated likewise by hopeful imaginaries for social and environmental justice (Stonich and Bailey 2000). The coalition responds to these imaginaries by pursuing the conditions of possibility for their future emergence.

Climate change poses serious threats to the coastal region of Bangladesh, as it does to all coastal communities around the world. Many of the impacts of climate change are already being experienced, and historically high levels of carbon emissions mean that they will continue to intensify. There is no question that coastal Bangladesh must grapple with the serious biophysical challenges presented by climate change (Bangladesh is not unique in this regard—significant, concerted climate action is a global imperative). The dynamics described here suggest ways in which the region and its inhabitants will be made dramatically more vulnerable to the impacts of climate change in the future. We see through them how power has shaped this landscape across multiple temporal and spatial scales. As we observed in chapter 1, these processes date to the colonial period. Today they continue to be shaped by decisions made by actors from the local to the global scale, from wealthy landholders in Khulna to policymakers at global climate negotiations. In order to understand the impacts of climate change, we must examine the interrelations between these historical patterns and ongoing processes of socioecological change.

The physical sciences have been able to successfully uncover and articulate the physical processes that are driving global climate change to a high degree of certainty (IPCC 2013). But the mechanisms of ecological transformation at the local scale are much hazier, and they are difficult to untangle from the role of global climate change. Attempts to attribute these local transformations to global dynamics have often marginalized their interactions with longer historical processes of change. In the process, they have naturalized historical and ongoing processes of dispossession. As this book has demonstrated, that marginalization can have serious impacts on how communities survive now and in the future. How we think about climate change shapes how we live with it.

In the process of enframing Khulna as a climate dystopia, the adaptation regime treats the dynamics of agrarian dispossession as externalities instead of understanding them as historical and systemic.[2] As we saw in chapters 3 and 4, research on ecological change in Khulna often obscures the normative politics of knowledge production and their role in shaping coastal ecologies. In so doing, it naturalizes and in turn reproduces this dispossession. In contrast, this book has demonstrated the need to understand the dynamics of dispossession in conjunction with climate change. I have investigated patterns of agrarian dispossession that existed prior to climate change and how they are reproduced through new imaginaries of climate crisis in the present.

Impacts of and responses to climate change are always mediated by power dynamics and histories in specific places. These changes are both structural, in the sense that they result from historical and existing power dynamics, and also contingent, in the sense that they are actively shaped and negotiated in the present. As Hall writes, "We make history, but on the basis of anterior conditions which are not of our making" (1985, 95).[3] The adaptation regime gives rise to new ways of governing communities and landscapes in coastal Bangladesh in the face of climate change, but it does so through existing ideologies and unequal power structures.

The unequal political economies into which climate change intervenes shape the way in which climate change is experienced by individuals and communities. Clearly, the negative impacts of climate change will be distributed unequally—climate crisis will be experienced most catastrophically by those who are already the most vulnerable. Yet this unequal distribution is not inevitable; it is the result of unequal modes of governing that predate this crisis and shape the way it will be governed in the future. The stories in this book highlight the importance of recognizing that the ways in which we understand and respond to climate change are fundamentally political. Ignoring the ways in which climate impacts are mediated by these existing politics both masks a complete understanding of the social and physical impacts of climate change and also risks exacerbating them. Moreover, inattention to the role of power dynamics in shaping communities and ecologies undermines opportunities for understanding how social movements can direct us toward new visions for climate justice.

The adaptation regime is a manifestation of these layered political ecologies. It obscures the impacts of climate change precisely by ignoring the histories and political economies that shape the way in which its impacts are experienced. In Bangladesh, the result is to entrench the unequal political economies that have emerged through centuries of colonialism and extractive neocolonial developmentalism that have already made rural communities profoundly vulnerable. Responding to climate change in this way is not inevitable, yet disrupting these unequal impacts of climate change would require confronting these histories directly. The adaptation regime is a response to climate change. In this sense, it is very specifically and concretely grounded in the present. Yet it is also profoundly shaped by these longer histories into which it intervenes. It is thus also deeply historical, reflecting power dynamics and forms of governing that long predate climate change.

Even as the adaptation regime is shaped by these existing political economic structures, it is also contingent, being actively shaped and negotiated by different groups of actors in the present. This contingency creates opportunities for transforming our political and economic systems even as it also creates opportunities for entrenching them. This study of the adaptation regime illuminates how this contingency can lead to further dispossession of already vulnerable communities, while simultaneously affording opportunities for resistance. Climate change and the discourses, practices, and interventions surrounding it mediate transformations in the political economy of development; some people will benefit from these changes while others will lose.

This is just one example of climate action leading to inequitable outcomes. There are many others. For instance, in the United States, response to disasters such as hurricanes has led to dramatic increases in the wealth gap between white and black communities, widening existing disparities between rich and poor (Howell and Elliott 2019). In other cases, the promotion of new renewable energy technologies (Borras and Franco 2018; Curley 2018; McCarthy 2015), carbon trading (Arora-Jonsson et al. 2016; Bond 2012; Osborne 2011), and other practices falling under the broad umbrella of "green capitalism" (Lohmann 2011; McMichael 2009; Prudham 2009) have led to the dispossession of land and resources, particularly from rural smallholders. In each example, the pattern and direction of redistribution are shaped by historical patterns and existing political and economic structures, but they also result from contested and highly contingent decisions made at the global, national, and local levels in the present.

Moreover, the ways in which we imagine and understand these landscapes and ecologies, and in particular the scientific tools we use to do so, both shape and are shaped by these political economies. It is necessary that we acknowledge not only future threats, but also how decisions are being made in the present by policymakers, financial actors, and others that are already actively transforming ecologies. Where our scientific analyses obscure the politics shaping socioenvironmental change and accept that the problems and solutions are inevitable, they foreclose opportunities for imagining and pursuing alternatives (O'Brien 2013). As we imagine the future with climate change, it is imperative that we open space for considering these alternative possibilities as new ways of pursuing climate justice. As we saw in chapters 5 and 6, communities and social movements in Bangladesh experience, negotiate, and contest these changes in different ways. In some cases, these contestations produce political visions that are radically more equitable than the dystopic climate imaginaries produced by the adaptation regime. These politics of possibility tell us not only that climate crisis is not inevitable, but also that better futures are possible.

My explorations of the simultaneously structured and contingent nature of climate action and climate impacts suggest that scholars interested in understanding climate change should direct their attention to historical and political economic context wherever they seek to understand socioecological change. In order to understand political alternatives, they must also look beyond expected sites of climate action (Koslov 2019). In another context, the adaptation regime would look quite different, mediating different power dynamics shaped by their own unique histories and ecologies. Thus, the adaptation regime is socially and historically specific, though its dynamics of *imagination, experimentation* and *dispossession* reflect patterns that can certainly be traced beyond Bangladesh. To analyze the workings of the adaptation regime in other contexts means considering these dynamics in light of the historical and political economic specificity of the unique ecologies and histories in those sites.

Imagination refers to the politics of knowledge production and anticipation of the future. As Anna Tsing writes, "Conjuring is always culturally specific, creating a magic show of peculiar meanings, symbols, and practices" (Tsing 2000, 119). Imagination reflects the imbrication of epistemological politics of the present with desires for the future. Even where certain imaginations of the future are hegemonic, they are not totalizing. The struggles against the dystopic imaginaries of the adaptation regime we observed in chapter 6 indicate the possibilities for resistance. Even where this resistance does not articulate itself in relation to climate change directly, it can suggest alternative ways of imagining a climate-changed future.

Experimentation refers to the provisional technological schemes that emerge in response to climate change. Like the adaptation regime that they serve, these experiments rarely address systemic power imbalances directly. By focusing responses on individual subjects of climate risk, these forms of experimentation divert attention from the political economic conditions that have given rise to climate change and the unequal manifestation of its impacts.

Dispossession is a consequence of the embeddedness of climate change and climate response in the political economy of capitalism. While dispossession is not an inevitable result of any response to climate change, any response to climate change that does not challenge the unequal political economy of capitalism will perpetuate and facilitate dispossession. In other contexts, the dispossession resulting from the adaptation regime may be less overt than that observed in Bangladesh, perhaps being mitigated by the kinds of social protections that impede the most extreme forms of capitalist exploitation in much of the world today.[4] Yet while it is possible to imagine ways of responding to climate change that resist dispossession, it is not possible to see climate change clearly without attending to the role of capitalism. The dispossession that shapes these responses to climate change is embedded in precisely the social and historical conditions that have produced climate change in the first place. Attempting to see climate change independent of these historical relations of capitalism would be to misunderstand the political economy of climate change itself.

Investigating the adaptation regime in specific social and historical conditions beyond Bangladesh will yield a different analysis of the political economies that mediate the impacts of climate change. I am hopeful about the possibilities for more equitable climate-changed futures. But writing capitalist dispossession out of this picture would make it incomplete. Dispossession in some contexts may be displaced onto other communities or ecologies, but it is inherent in the social relations of climate change as an ecological manifestation of capitalism.

The framework I have outlined here for understanding the adaptation regime thus contains general patterns that will no doubt be identifiable in other contexts. The adaptation regime is a global phenomenon, containing forms of governance that manifest at multiple scales. I have also demonstrated in this book the imperative of understanding this regime of governance through specific sites and the ways in which their unique histories and relations of power shape climate response. Any general theory of adaptation not attuned to this specificity misses an essential point: that the governance of climate change adaptation is intimately linked with existing governance regimes and the power relations under climate change are mediated by existing power structures. It is only through interrogating those existing power structures that we develop a workable account of the power relations that shape climate change adaptation and its outcomes.

Centering the specificity of power relations in particular places not only offers us a better accounting of the contemporary governance of adaptation, it also opens up novel possibilities for imagining more just climate futures. As the stories contained in this book highlight, imagining climate futures on different scales exposes radically different political possibilities. On the global scale, imaginaries of the future of coastal Bangladesh suggest the inevitability of crisis. Being divorced from the imaginaries produced within these communities themselves, the adaptation regime promotes visions of the future that limit the ability of alternative imaginaries and futures to emerge.

The impacts of climate change are produced and refracted through a multitude of temporal and geographic scales. Bangladesh is not more vulnerable to climate change than other communities around the world because of its "bad latitude" (Watts 2003). Rather, Bangladesh is more vulnerable because of a planned historical process of development within the global capitalist system. To understand climate change as somehow transpiring outside this history disavows its formative linkages with colonial and capitalist exploitation (Ahuja 2016; Whyte 2017a, 2020). Indeed, such discourse of climate change, in the words of Michael Watts, "feeds the great semiotic machine that naturalizes the consequences of social exploitation" (2001, 139). Hence, to examine climate change in relation to these planned and intentional processes is to denaturalize this exploitation and question its inevitability.

The impacts of climate change have never been inevitable.[5] There is still much that can be done through the abandonment of fossil fuels to mitigate the worst effects of climate change. Today, at the global scale, existing and projected global warming is the result of particular decisions about industrial growth and resulting emissions calculated against known effects (Rogelji et al. 2016). Likewise, at the local scale, in every community, choices are being made now to address these changes that also drive ecological transformation in the future. A recognition of this human agency and its histories is the first step in forging a radical politics of climate justice.

Methodological Appendix

In the Introduction, I provided a brief overview of how I came to this project through my work with Nijera Kori. Here I provide a closer look at the methodological choices I've made over the course of the research carried out for this book, why I made them, and the collaborators and interlocutors who helped to shape what it became. In addition to providing more detail on the methods used to conduct this research, this Appendix also offers insight into my particular subject position vis-à-vis various actors represented in the book and how this positionality has shaped the conclusions represented here.

While my fieldwork for this project began in rural communities in Khulna in 2012, when I moved to Bangladesh in 2014 to begin two years of concentrated research, I began to reach out to the development community in Dhaka. I reached out to and conducted interviews and participant observation with donors, development practitioners, Bangladeshi government officials, and scientists. These actors make up the community that I came to think of as Bangladesh's adaptation regime. This included key decision makers in major development programs at every major aid agency working in Khulna and Bangladesh government agencies that was concerned with development planning in Khulna. The activity among these various actors in Dhaka is happening constantly, which often means that there are several events in a week to which key actors are invited and at which relevant agencies are expected to be represented. An important part of the work of understanding this community, then, was to attend all the relevant seminars and workshops, get to know all the key players at different agencies and embassies, and become immersed in the everyday practices of the community. To the extent that I was able, I attended these events regularly in order to become part of the community myself and participate in its ongoing dialogues. I also conducted private interviews with many of the participants whom I met through these activities, usually in their offices or cafes in Dhaka that were frequented by expatriates. I obtained verbal permissions from all interviewees beforehand.

Becoming immersed in this community of development actors often posed methodological and ethical challenges related to working within communities holding often radically different political commitments and epistemological perspectives. I confronted particular challenges related to the antagonistic relationship of activists and development actors, particularly in relation to conflicts over shrimp aquaculture. After years of working with activists in Bangladesh, I was repeatedly surprised to hear derision of them among development actors who seemed to assume I would share their perspectives. I was careful to disclose to development actors in Dhaka my ongoing work with activists in Khulna when it was relevant, while also not letting these relationships overdetermine the content of conversations with actors with other viewpoints.

A common event among development agencies is the "Stakeholder Workshop." While this title might suggest the attendance of those rural community members who are the subjects (or objects) of development programs, the Stakeholder Workshop is instead an event to which development practitioners and government bureaucrats are invited. I was also frequently able to talk my way into invitations to such events. I was often very aware that my presence as a participant observer at these events may have been construed as consent to the program proposals under discussion. Attendees at these events are commonly offered an envelope of cash as a per diem or "conveyance" (theoretically a transportation stipend) in recognition of their attendance. I observed that these payments were sometimes as much as 5,000 Bangladeshi taka (about US $60) among larger aid agencies. I did not accept these cash payments in order to indicate my role of observation as opposed to endorsement.

My access to this community of practice concerned with climate change adaptation was expanded about nine months into my fieldwork when I was invited to become a visiting researcher at the International Centre for Climate Change and Development (ICCCAD) in Dhaka. ICCCAD's director, Dr. Saleemul Huq, is a leader in global climate change policy and diplomacy networks. Because of the organization's leadership in these conversations within Bangladesh, ICCCAD is a key node in the adaptation regime. Dr. Huq regularly brought me with him to high-level seminars and meetings to which he was invited and facilitated exceptional access to the inner workings of these networks and conversations. My relationship with ICCCAD gave me new insight into the threats climate change poses to Bangladesh and the international negotiations surrounding climate treaties. These insights also gave me new sympathy for the political claims made within these discourses and helped me to reframe my analysis of what climate justice could mean for communities in coastal Bangladesh and beyond.

In addition to working with these development practitioners, I also carried out interviews and participant observation with a variety of scientists and researchers who study dynamics of environmental change in coastal Bangladesh. This includes both Bangladeshis and also foreign researchers primarily from the United States and United Kingdom. I have included more information about my relationships with these researchers in chapters 3 and 4, so they do not require extensive elaboration here. Because of their funding, their collaborations, and the conditions through which many of these researchers come to work in Bangladesh, I think of them as liminal actors in the adaptation regime.

My research in Khulna built directly on relationships I developed and research I had conducted previously alongside Nijera Kori. In 2012, I visited Khulna for the first time with Rezanur Rahman Rose, Nijera Kori's director of research, who is a longtime activist and community organizer with the organization. Rose has been a trusted colleague and interlocutor for over a decade, and I selected my research sites collaboratively with him. In 2013, I returned to these sites to conduct a study of the agrarian political economy of shrimp aquaculture with my longtime collaborator Jason Cons and members of the Nijera Kori landless collectives (Paprocki and Cons 2014). To do this research, we used a participatory method we call Community-Based Oral Testimony (CBOT), which we had used previously to study community perceptions of microcredit in northern Bangladesh (Cons and Paprocki 2010; Paprocki 2016). This research provided a robust foundation for my ethnographic study of related dynamics in Khulna, and it was a significant benefit to have already established relationships in each of these communities.

The three field sites in Khulna in Polders 22, 23, and 29 highlight how the impacts of environmental change are being experienced in different ways. I chose them because they have taken very different approaches to navigating ongoing transformations in the region and also because they illuminate different possibilities for ongoing rural livelihoods there. The first community is in Polder 23, which has almost entirely transitioned from rice agriculture to shrimp aquaculture (starting in the 1980s). Across the river, the community in Polder 22 has historically resisted the transition to shrimp largely through the mobilization of a landless social movement supported by Nijera Kori. Farmers in the community in Polder 29 have transitioned from shrimp back to the earlier mode of agricultural production through the same social mobilization seen in Polder 29 (also through the support of Nijera Kori community organizers). I regularly traveled by bus between Khulna and Dhaka in order to continue ongoing work between these varied communities. While this approach required some compromise in the continuity of ethnographic observation in both Khulna and Dhaka, it had the benefit of allowing me to observe the development and impacts of the adaptation regime as both multiscalar and contingent. In Khulna this approach also had the benefit of allowing me to observe community dynamics through different growing seasons for shrimp, rice, and other agricultural crops.

While traveling around rural Khulna, I stayed at the remote field offices where Nijera Kori staff live and I frequently spent time with Nijera Kori members and local leaders. On occasion, movement members would ask to be interviewed within Nijera Kori compounds, which are generally relatively private compared to domestic spaces in rural communities (such as homes or gardens, where neighbors will often wander in and crowd around conversations uninvited). However, I did not let Nijera Kori staff accompany me while I was walking around the villages conducting interviews. This, I believe, allowed those community members who were not in the movement to feel more free to discuss their perspectives with me. Having already spent time in each of these villages during the prior two summers, by the time I began the dedicated period of fieldwork for this book in 2014, most of the residents had heard of my presence and my relationship with the movement. Yet, perhaps because of the significance (in different ways) of shrimp aquaculture in each of these communities, most were excited to relay their own perspectives and personal stories, regardless of their own relationship with shrimp production. Their awareness of my relationship with Nijera Kori, I believe, also led them to a desire to account more thoroughly for their motivations in participating in the shrimp boom, details that enriched and nuanced my research findings.

In Polder 23, I noticed that the out-migration of many community members meant that I was missing the perspective of a significant portion of the population. This migration is significant both because of what it tells us about dispossession and local agrarian political economy and also because of widely circulating narratives about climate migration from Khulna. Since respondents there frequently told me stories of friends or family members who had migrated to a particular slum on the outskirts of Kolkata, I began to solicit mobile phone numbers of these migrants with the intention of traveling to Kolkata to find and interview them. The vast majority of the phone numbers I gathered did not lead to viable interviews, either because a migrant's prepaid phone was out of service or perhaps because of the discomfort of some of the people I contacted with being interviewed by an unknown stranger. I intentionally did not solicit or document information about these migrants' legal status in India, though presumably many if not all lacked the legal authorization to work or reside there. I assume that their legal status made many concerned about speaking with me, their fears heightened by a growing xenophobic rhetoric in the midst of the 2014 Indian elections, during which then-candidate Narendra Modi remarked, "You can write it down. After [the election on] May 16th, these Bangladeshis better be prepared with their bags packed" (*Dhaka Tribune* 2014a). Two of the phone numbers that were shared with me in Polder 23 led to interviews with young men who had migrated from there, and these interviews led to a snowball sample of approximately twenty interviews with other migrants living in the same slum. I see this work as an extension of my research in the field site in Polder 23.

I also conducted research internationally at conferences on climate change and adaptation in Paris, Rotterdam, Kolkata, Kuala Lumpur, and Bonn. This research helped me to better situate the dynamics of the adaptation regime I was observing in Bangladesh within a broader global context and to understand Bangladesh's role within this global regime. In Kolkata, this participation also led to opportunities to conduct interviews with Indian NGOs and government officials conducting work in the Sundarban mangrove forest on the other side of the border from Bangladesh. This work extended my understanding of the cross-border environmental planning and interventions and their relationships to linked but distinct development regimes.

Finally, I conducted historical research in archives in Dhaka, Khulna, Kolkata, Delhi, London, Wageningen, Washington, DC, and outside Hamburg. This included government, NGO, and university archives, as well as archives in the private collection of Gertrud and Helmut Denzau, a German couple who, as amateur naturalists with a passionate interest in the Sundarbans, have amassed what they (I believe rightly) consider to be the most comprehensive scholarly and historical archive on the Sundarbans in the world in their home in a small village near the Baltic Sea. Collectively this archival research facilitated an understanding of the history of human intervention in this region and how this intervention has shaped the current landscape and the vulnerability of communities inhabiting it.

Glossary

***aman* rice**—Bangladesh's most important rice crop, which is cultivated during and after the monsoon.

aquaculture—Controlled cultivation of aquatic animals in penned enclosures, often on former farmlands.

badh—An embankment.

***bagda* shrimp**—Giant tiger shrimp that survive in saltwater, existing in the wild and also cultivated in southern Bangladesh.

beel—A marshy lowland.

bigha—A variable unit of land area commonly used to measure size of farming plots. In most of Khulna, one *bigha* is equivalent to approximately 33 decimals, or one third of an acre.

boro lok—Literally "big guy." A popular colloquialism used to refer to elites with economic and/or political power.

gher—An enclosed area of land flooded with salt water to cultivate shrimp.

***khas* land**—Government lands legally available for access among landless populations according to Bangladesh's constitution.

maund—A local unit of weight standardized throughout South Asia under the British colonial government to 37.3242 kilograms. *Maund* per *bigha* is the standard unit by which yields are measured at the farm level in Khulna.

nosimon—A three-wheeled vehicle built by local mechanics using small, diesel-powered tractor engines, used for transporting people and goods.

oshto masher badh—"Eight month embankment." An indigenous system of management used historically in which embankments were built and deconstructed seasonally to facilitate agricultural production.

polder—Dutch word for a low-lying enclosure surrounded by protective dykes.

pucca—Ripe, mature, permanent. Used to refer to homes made of brick or roads paved with asphalt.

sheola kaj—Literally "algae work." Labor conducted primarily by women in shrimp ghers to remove scum from the surface of the water.

shrimp fry—Shrimp at the postlarvae stage. A term used primarily by the aquaculture industry to refer to the "seed" used to stock a gher for raising shrimp. Fry can be caught in the wild or cultivated in hatcheries.

taka—The Bangladeshi unit of currency (denoted by the symbol ৳ or Tk). $1 USD is equivalent to approximately 85 Tk.

thana—Term for a police station, but also referring to an area controlled by a police station; formerly the term for the administrative subdistricts now known as *upazilas*.

Union Parishad—The smallest rural administrative unit in Bangladesh; each union is made up of nine wards, usually each ward contains a single village.

upazila—An administrative subdistrict; formerly known as a *thana*.

zamindar—Traditional large landowners in Bengal, who were responsible for revenue collections under the Permanent Settlement laws of British colonial administration; today a *zamindar bari* is the former home of a *zamindar*.

Notes

INTRODUCTION

1. My use of "enframing" draws on Mitchell (1988) and is also informed by his epistemological analysis of "externalities" (2002).

2. Most of the donors represented in this book are from the United States, Canada, and the countries of the European Union. However, I also interviewed donors from Japan, South Korea, and the Asian Development Bank, as well as individuals at international agencies such as the World Bank, which employs an international staff representing many countries of the Global North and Global South.

3. See Zeiderman (2016) for an examination of anticipatory modes of governing in the context of climate change in Colombia.

4. In both cases in New York, climate change plays a minor role in the discourse of threat and adaptation, particularly in comparison to Bangladesh, where it becomes a major operating logic in the political economy of development. This limited recognition of the threat of climate change to New York City itself reflects discrepancies of imaginaries of which communities are threatened by climate change and which are not.

5. The agrarian question has always revolved around the drivers and trajectories of the transition to capitalist production relations and the impact on the agrarian classes. While there has been a great deal of debate about the nature of twenty-first-century agrarian change (Bernstein 2016; McMichael 2006; Watts 2002), I suggest here that climate change (and responses to it) are the most significant drivers of agrarian change in the twenty-first century (see also Paprocki 2020).

6. In contrast, Whyte has argued that some indigenous communities imagine the Anthropocene as already a kind of dystopic future (2017b, 2018).

7. This phrase, which is representative of climate discourse throughout Bangladesh's development and donor community, appeared in 2014 in the third newsletter of the Bangladesh Climate Change Resilience Fund, a now-defunct multidonor trust fund administered between 2010 and 2017. I expand on this discourse in chapter 2.

8. I thank Dina Siddiqi for this point about child marriage, which she made in a presentation at the 2017 Association for Asian Studies annual conference.

9. All translations in the manuscript are my own.

10. In this sense, the book draws on a long history in agrarian studies of insisting that history *matters* (Akram-Lodhi, Borras, and Kay 2006; Wolford 2007).

11. Though Bangladesh's NGO sector is today dominated by organizations focused on microcredit and other service delivery, a "radical NGO sub-sector" emerged in the aftermath of the country's independence. Focused on consciousness-raising particularly among the agrarian poor and nonparty political mobilization, this sector has declined due to a variety of institutional and political factors operating at local, national, and international scales (Lewis 2017). Nevertheless, this use of "NGO" to signal apolitical service delivery work is a functional but perhaps oversimplified generalization of a diverse and unstable category of organization (Lewis and Schuller 2017).

12. In 1972, the Indian government unilaterally opened this barrage roughly ten miles upstream from the Bangladesh border. Though its impacts have been much debated, the

barrage has been linked with serious ecological impacts in Bangladesh, particularly in Khulna (Adnan 2009; Brown and Nicholls 2015; Kimberley Anh Thomas 2017a).

1. "SLUTTISH, CARELESS, ROTTING ABUNDANCE"

1. In the colonial record, the Sundarbans were variably referred to as "Sunderbunds," "Sunderbans," "Soonderbans," and so on. This variability is largely the result of an uncertain etymology, which attributed the region's name in turns to the beauty of the forest, its proximity to the sea, the presence of embankments, and the abundance of sundori trees (a particularly valuable timber resource), among other theories. Unless quoting directly, I used "Sundarbans" to reflect the accepted current translation and transliteration. The contemporary district of Khulna in southwestern Bangladesh (on which this study is focused) falls within the broader area historically referred to as the Sundarbans. In 1927 the colonial settlement officer of Khulna wrote that "the revenue history of the area covered by Khulna district is almost entirely a history of Sunderban administration" (Fawcus 1927, 64). The bounds of this region have shifted over time, both as a result of deforestation that expanded the area of human settlement, changing borders of districts and nation-states, and ongoing transformations in juridicopolitical regimes. The Khulna district itself was not officially demarcated until 1882 (Heinig 1892, E2). On the Indian side of the border, the name "Sundarbans" refers to the broader region including the inhabited islands outside of the protected mangrove forests. In Bangladesh today, it refers instead only to the protected areas, and not the now-deforested surrounding islands. Thus, my use of this variable term is complicated by its unstable application across the border, and it is further complicated temporally by the deforestation of the mangroves throughout this time period (peaking during the colonial period). In this chapter I thus take the historical region of the Sundarbans, however unstable, as a rough proxy for the region examined in this study more broadly. In later chapters, my focus is on agricultural landscapes that inhabit these deforested and reclaimed forest tracts.

2. Supported by the East India Company, Major James Rennell carried out the first comprehensive survey of the Ganges, Meghna and Brahmaputra River systems in the late eighteenth century. His *Bengal Atlas* and accompanying *Memoir of a Map of Hindoostan* were essential texts employed by the British in their conquest and management of the subcontinent (Barrow 2003; Rennell 1781, 1788).

3. I begin with the colonial period not to deny that the political ecology of the region was shaped in important ways prior to British intervention. (See Sivaramakrishnan 1999 for a critique of the preoccupation among postcolonial scholars with colonialism as a watershed in South Asian environmental history.) Indeed, according to the Bangladesh Bureau of Statistics, the area of Paikgachha was settled under Hazrat Khan Jahan Ali, a legendary fifteenth-century Sufi saint and governor of what is now the Khulna region under the Bengal Sultanate (BBS 2012). The name "Paikgachha" is said to be a reference to a particular group of peasant-militias drafted into service for clearing the Sundarbans (the *paiks*) and the trees they were forced to clear to make way for cultivation (*gach*).

4. Richard Eaton has examined precolonial jungle reclamation and the advance of the agrarian frontier in this region, particularly the role of Muslim holy men in the early Mughal period (1990, 1993). While the time frame of early settlement in the region is the subject of much debate (Beveridge 1876b), some scholars have dated settlement in the Sundarbans region as far back as the Ramayana and Mahabharata religious texts (around 3000 to 4000 B.C.) (Chattopadhyaya 1999, 26).

5. A distributary is a smaller stream that branches off from the main channel of a river (and thus is the opposite of a tributary). Distributaries are a common feature of river deltas.

6. Allison et al. cite a rate of avulsion of the major rivers in the region—meaning complete abandonment of one river channel and formation of another—at the time scale of a single century (2003, 319).

7. Though the interaction between natural and anthropogenic factors driving climate change is poorly understood, Brown and Nicholls documented the human influences in the Bengal Delta and explained that the most significant driver of anthropogenic subsidence in Khulna has been reduced sedimentation caused by protective embankments (2015). A thorough examination of anthropogenic drivers of subsidence in delta regions can be found in Syvitski et al. 2009.

8. See also Camargo 2017, which examines similar dynamics in a Colombian floodplain in the context of climate change adaptation.

9. This post of commissioner of the Sundarbans existed until 1905, when the Sundarbans Act transferred the powers of the commissioner to the collectors of the three districts within which the forest was contained (Khulna, Backarganj, and the 24-Parganas) (Mandal 2003). By this time the Sundarbans had been so comprehensively mapped and surveyed that the land administration was not so burdensome as to require its own administrator.

10. The task of determining a precise definition of "wasteland" is, however, confounded by the very mutability of its application (Baka 2013). Waste is a relational concept, which is constantly redefined alongside shifting development objectives and theories of value (Baka 2017; Gidwani 2012; Harms and Baird 2014).

11. Wasteland discourses have come under renewed scrutiny in recent years due to their revitalization within the rapid growth of large-scale land deals (the "global land grab") (Borras and Franco 2012; Ben White et al. 2012; Wolford et al. 2013). Much of this literature has examined the roots of current land-grabbing dynamics in the colonial enclosure movement (Baka 2016; Ben White and Dasgupta 2010). Historians of South Asia have likewise recognized the significance of wasteland development schemes and narratives in the British colonization of the subcontinent (Gilmartin 2003; Iqbal 2010; Ludden 2011).

12. Sivaramakrishnan explains that the establishment of the Bengal Wasteland Rules in 1853 served a variety of purposes for the colonial administration, which were not always directly related to the extraction of forest resources (1999, 132). The present case of wasteland categorization supporting the reclamation of land for rice cultivation supports this analysis.

13. These leases were known as *patitabad*, which can be translated as "fallen settlement" or "degraded settlement," reflecting the sense of these lands as offering exceptional potential for exploitation. The term also reflects the moral valence of "wasteland" classification, which is explored in more detail by Gidwani (1992) and Baka (2013). I am grateful to K. Sivaramakrishnan for his insight into the Bengali and Urdu etymology of this term.

14. As a counterpoint, it is worth noting here that historian P. J. Marshall attributes the expansion of cultivation in the Sundarbans primarily to "climatic regularity rather than colonial rule" (1987, 180).

15. After gaining independence from the British in 1947, India was partitioned into two separate states, India and Pakistan. Pakistan was composed of two physically separated geographic bodies, West Pakistan (the contemporary state of Pakistan) and East Pakistan (contemporary Bangladesh). The relationship between these two exclaves was always tenuous, as it was organized uneasily around a shared Muslim-majority population and characterized by intense cultural, linguistic, ethnic, and economic diversity.

16. Rainey himself was a charismatic and infamous character, who purchased and inhabited a large *zamindari* estate near Khulna, which he called "Rainey Villa" (Rainey 1897), for the purposes of cultivating indigo (Westland 1874). In 1874, Westland, who

was then the magistrate of Jessore, wrote that the subdivision of Khulna had been originally established in 1842 with the "chief object . . . to hold in check Mr. Rainey, who had purchased a zemindari in the vicinity and resided at Nihalpur, and who did not seem inclined to acknowledge the restraints of the law" (221–222). The antipathy between Rainey and colonial administrators appears to have been mutual. He was among the most vociferous opponents of the transition away from the promotion of Sundarban land reclamation (examined further later in the chapter), which he described as "retrograde" (Rainey 1891, 279), accusing the government of "neglect" for failing to recognize the role that Sundarban reclamation could play in the mitigation of famines (Rainey 1874, 332).

17. In the colonial period, both the British and the Bengalis consistently referred to the Arakanese, a group which is indigenous to a shifting region straddling what is now the border of Bangladesh and Myanmar, as "Mugh" (Eales 1892, 197). I avoid using the term here, however, because it is considered to be derogatory.

18. This myth of the "lazy native" was common across colonial South and Southeast Asia (see also Peluso 1992).

19. The history of habitation in this region is contested, but Mukerjee asserts that land reclamation began in the region in the fifteenth century (Mukerjee 1938, 137), while Eaton suggests that forest clearing for the intensification of wetland rice agriculture began here in the thirteenth century under the rule of independent Indo-Turkish sultanates (Eaton 1990).

20. "Premature reclamation" is a contested category. While it refers to the act of constructing embankments to artificially trap sediment and build up land, there is no obvious boundary between this kind of intervention and the act of constructing an embankment around a secure piece of land to prevent erosion and inundation. Given the active nature of the Delta, in the absence of intervention, any piece of land relatively proximate to a river is liable to erode away in due course.

21. Subsidence is also the result of tectonic movement and sediment compaction. However, the net result of these physical processes was historically counteracted throughout most (though not all) of the coastal region by ongoing tidal sediment deposition (Nicholls et al. 2013; Wilson and Goodbred 2015). Thus, this current observation of change in elevation is most appropriately described as net subsidence.

22. One USAID report explicitly states that "the scope of the project was expanded in fiscal year 1962, on the basis of a plan prepared by a U.S. engineering firm" (US General Accounting Office 1971, 55), suggesting that Leedshill–De Leuw was initially hired to manage the CEP and that the firm's management quickly expanded the scope of the work it was intended to undertake.

23. Thomas cites similar problems in 1964 and 1965 with another EPWAPDA embankment project further north (the Ganges-Kobadak Project). Out of desperation for water to irrigate their rice fields, farmers cut the embankments to allow water to flow in from the river. On both occasions they were fired on by the police and several farmers died (John W. Thomas 1972b, 16).

24. The maintenance of water infrastructure is technically within the mandate of the Bangladesh Water Development Board, although the capacity for this maintenance is not provided for in dramatically scaled-up infrastructure programs, and it often falls short.

25. Reclamation through large-scale grants of lands in the Sundarbans was officially discontinued in 1903 (Iqbal 2010, 27).

26. The legacy of this agricultural labor migration can be observed in the current humanitarian crisis of the Rohingya, a stateless minority descended from landless agricultural laborers who migrated in the period prior to Partition from the region that is now Bangladesh to the region that is now Myanmar (Bjornberg 2016).

27. In 1871, the commissioner of the Sundarbans reported that there were 431 such temporarily settled estates (Hunter 1875a).

28. WorldFish, for example, is headquartered in Malaysia, while IRRI, the International Rice Research Institute, is based in the Philippines.

2. THREATENING DYSTOPIAS

1. This chapter is derived, in part, from an article published in *Annals of the American Association of Geographers* 108 (4) (2018): 955–973 and available online at https://www.tandfonline.com/doi/full/10.1080/24694452.2017.1406330.

2. DFID was closed in 2020 and replaced by the Foreign, Commonwealth & Development Office.

3. These practices have been a focus of political ecology since the publication of Piers Blaikie's *The Political Ecology of Soil Erosion in Developing Countries* (1985).

4. The "basket case" comment has regularly been attributed to Henry Kissinger, although Lewis instead credits Ural Alexis Johnson, then a US undersecretary of state for political affairs (Lewis 2011; see also Sarah White 1999). At the time the comment was specifically in reference to the war-torn country's impending famine, though the epithet came to take on a variety of derogatory connotations about Bangladesh as underdeveloped, overpopulated, and lacking in self-sufficiency.

5. A transcript of a 1971 phone call between US president Richard Nixon and his secretary of state, Henry Kissinger, highlights these early tensions. Following the receipt of a telegram from the US Embassy in Dhaka titled "Selective Genocide" and concerning violence against Bengali civilians by the Pakistani military, Nixon expressed his support for General Yahya Khan, the president of Pakistan, noting that "the real question is whether anybody can run the god-damn place," to which Kissinger responded, "That's right and of course the Bengalis have been extremely difficult to govern throughout their history" (Blood, Nixon and Kissinger 2013:243).

6. The word *thana* refers to a unit of police administration, which today are known as *upazilas*, or subdistricts.

7. Not all appraisals of BARD's legacy are so bleak. Lewis (2019) for example, regards Khan as an early pioneer of NGO development focused on grassroots perspectives and the importance of autonomous community-level organization.

8. In his historical study of PARD, Ali found documentation of extensive complaints from women with these contraceptive technologies, and particularly side effects from IUDs, which included "heavy bleeding, irregular menstruation, abdominal pain and weakness" (Choudhury 1969, cited in Ali 2019, 439). These side effects were attributed by PARD to women's health conditions, not to flaws with the contraceptive technologies.

9. On the performance of vulnerability in service of climate change adaptation, see Farbotko 2010a; Haalboom and Natcher 2012; and Webber 2013.

10. As a literary genre, these memoirs share similarities with what Lewis (2014) has referred to as the "development blockbuster," an autobiographical genre of books published by commercial presses concerned with conveying expert knowledge of the author that unsettles conventional development wisdom.

11. E.g., Kroodsma 2015; Mifflin 2013; Oxfam 2010; Voysey 2015.

12. Bangladesh is located in South Asia.

13. See also Bassett and Fogelman (2013).

14. On the Climate Smart House as a spectacle of securitized visions of climate futures, see Cons (2018).

15. In 2015, 66 percent of the population was rural, down from 95 percent in 1960 (World Bank 2017).

16. These findings are confirmed by my own research, as well as by Datta (2006), among whose respondents 90.7 percent cited decreases in vegetation due to shrimp cultivation.

17. This celebration draws glaring parallels with earlier dynamics of agrarian dispossession in moments of political economic transition. See also Federici (2004).

3. OPPORTUNITY/CRISIS

1. For a more extended discussion of the particular agrarian political economy in Dacope and mobilization against shrimp there, see Afroz, Cramb, and Grünbühel (2017).

2. What is known, however, is that these dynamics are embedded in broader secular dynamics of social and biophysical change in the Delta, which is in an active state of dynamic transformation.

3. There is a great deal of uncertainty and contestation about the effects of the Farakka Barrage, with analyses of its physical impacts being fairly firmly divided between India and Bangladesh. The Indian government, public, and scientific community largely dispute the impacts claimed by their corresponding communities in Bangladesh. India built the barrage initially in order to bring more water to the Port of Calcutta, which was experiencing excessive siltation due to decreased flows, obstructing the traffic of ships to the port.

4. Rates of erosion and accretion are, however, variable across Bangladesh's coastal zone, and in some areas, erosion outpaces accretion (Sarwar and Woodroffe 2013).

5. Goldman has described similar dynamics of liminality in relation to what he calls "hybrid state actors" (2005, 38). Similarly, in her study of the US Department of the Interior, Megan Black examines the creation of a cadre of Third World scientists, which is invested both in the technologies of imperial extraction and in the development funding that accompanies them (2018).

6. For more on unequal relations of power in North-South research networks, see also Landau (2012). Landau in particular highlights how funding constraints shape these unequal research partnerships in ways that often push southern researchers toward policy-oriented research that constrains scholarly engagement and fundamental critique. These dynamics can also clearly be observed in Bangladesh.

7. One World Bank administrator explained to me later that this was the result of a legal technicality in the World Bank's procurement policies.

8. Given the ambiguity between "foreign" and "local" forms of expertise discussed earlier in the chapter, this seems to me to be an open (but nevertheless critical) question. All the faculty at IWFM are of Bangladeshi origin yet most of them earned their advanced degrees abroad, primarily in the Netherlands, so their own knowledge and expertise are fundamentally transnational.

9. Community-Based Adaptation (CBA) is a particular field of practice involving its own series of publications, transnational networks of practitioners and researchers, and conferences (the first of which was held in Bangladesh in 2005). Forsyth sees CBA as part of a broader trend to link climate policy with international development practice (Forsyth 2013). In this sense, CBA plays an important role in the adaptation regime.

4. THE SOCIAL LIFE OF CLIMATE SCIENCE

1. I refer here to Foucault's usage of episteme as "the 'apparatus' that makes possible the separation, not of the true from the false, but of what may from what may not be characterized as scientific" (1980, 197).

2. There is also evidence that the rapid deforestation carried out to allow for shrimp farming has contributed to increased erosion along the river banks, resulting in an increased sediment load in the rivers and increased siltation in the distributary channels around the polders (Deb 1998).

3. There are several words in the Bengali language that are used to refer to marshy lowlands or wetlands, including *beel*, *haor*, *baor*, and *jheel*. There is little technical difference between them (although in practice they sometimes differ in size or in regional usage). *Beel* is the word that is primarily used to refer to such lowlands in southwestern Bangladesh.

4. The first major attempt at this translation was through the Khulna-Jessore Drainage Rehabilitation Project (KJDRP), a major infrastructural program financed by the Asian Development Bank (ADB) to improve drainage infrastructure and increase agricultural production. Due largely to these failures of translation, the KJDRP was found to have been unsuccessful by activists and the ADB alike (Asian Development Bank 2007; Shahidul Islam and Kibria 2006).

5. These tensions over attribution of the causality and responsibility for social and ecological damage caused by shrimp aquaculture are also reflected in Béné's analysis of the epistemological politics that are embedded in global discourses of shrimp aquaculture development (Béné 2005). Béné explains that those for and those against shrimp aquaculture differ in their focus on the problem's causes rather than its solutions and also on the respective responsibilities of small- and large-scale farmers, local politics, development agencies, and environmental NGOs. He also found that technical experts promoting shrimp aquaculture place much of the blame for unsustainable conditions of shrimp aquaculture on NGOs for their lack of technical knowledge, misunderstandings, and spreading of rumors about the impacts of shrimp aquaculture. He added that technical experts dismiss the knowledge claims of these activists and NGOs as "emotional" and "unscientific." These conclusions are in line with my own research findings, particularly the frequency with which I heard aquaculture experts privately ridicule activists and NGOs for what they described as a lack of information and understanding of technical developments in the aquaculture field. See the further analysis of these epistemological politics in chapter 5.

6. There is a robust critical literature in Development Studies on the "logframe" or "logical framework" approach, a methodology used by international development agencies for designing, monitoring, and evaluating development projects (Gasper 2000; Kerr 2008; Prinsen and Nijhof 2015). It is presented as a table, within which the rows list activities, outputs, purposes, and goals and the columns list project description, indicators, means of verification, and assumptions. It aims to convey a sense of rational and goal-oriented project planning. Lewis writes that the use of the logframe as a planning tool serves to divert attention from the political and historical context in which development programs intervene (2010). The origins of the logframe approach lie in US mid-century military planning, which focused on strong central authority, streamlined goal setting, and quantifiable objectives.

7. In this sense, these maps share some epistemological terrain with the rate maps for flood insurance issued by the US National Flood Insurance Program (as described to Elliott [2017]) in the sense that they attempt to concretize a fundamentally moral (as opposed to physical) approach to land management.

8. This is likely for the first organizations listed but less so for the independent researchers; however, academics are often hired as individuals or departments for private consulting work, and in that context, such academic analysis can be used for project design.

5. AUTOPSY OF A VILLAGE

1. Elsewhere in Polder 23, elderly residents reported to me having grown up to three crops in a year, including an additional crop of *aus* rice, a shorter, dry-season rice grown between the dry and monsoon seasons. *Aus* is typically grown on higher land, so it is less likely that it was ever grown in Kolanihat, much of which is on relatively lower ground.

2. Afroz, Cramb, and Grünbühel (2016), Datta (1998), and Talchabhadel et al. (2018) documented the same system being used historically in neighboring communities in Khulna.

3. While it is often said that the alluvial soil deposits are nutrient-rich, thus facilitating this fertilization, Roy, Hanlon, and Hulme write that the increased fertility is derived from nutrient-rich algae breeding on the flooded land, which produces natural nitrogen fertilizer (2016, 44).

4. Afroz, Cramb and Grünbühel attribute the breakdown of this system in Dacope to the construction of the polders in the 1960s (2016). Thus, the relative roles of the polders versus the breakdown of the *zamindari* system in the collapse of this indigenous water management system remain an open question.

5. Units of land measurement have been converted from the local unit of *bigha*, which is approximately one third of an acre.

6. The Bengali word used to describe these guards hired from the outside is *mastan*, which translates to "goon," "thug," or "mobster."

7. Wajed Ali is a well-known shrimp businessman and notorious land grabber in Khulna. I have retained his actual name here due to his public notoriety. He will appear again in the following chapter.

8. Hydrological units are spatially, not politically delimited, with boundaries set by large embankments or major roads, and thus often pass through the borders of a village or other political unit.

9. As discussed in chapter 1, the polder embankments were constructed by the East Pakistan Water and Power Development Authority (EPWAPDA). The Bengali word for these major external embankments (*badh*) is the same as the word used for small earthen embankments between fields. As a result, in Khulna the embankments surrounding the polders are often referred to as "WAPDA *badh*," or just "WAPDA" for short.

10. Some also report that the water has iron in it, which could indicate a more serious contamination. The presence of iron is visible as it leaves red deposits on the ground when water evaporates. Iron also frequently co-occurs with arsenic, which is a major problem with tube wells in Bangladesh. As arsenic is invisible, people are more likely to detect a contamination with iron than they are with arsenic.

11. These bunches are sold by the *maund*, which is equal to approximately 37 kg.

12. See also Paprocki and Cons (2014) on the collapse of common-pool resources in Polder 23.

13. *Zamindar baris* were the residential palaces and centers of business administration of the large colonial-era estates of *zamindars*. Many were abandoned after the abolition of the *zamindari* system in Bengal after Partition. Today some of the larger estates have been turned into museums, while others are inhabited by squatters or the descendants of their former proprietors.

14. Shrimp cultivation is certainly not the only cause of the change in this family's fortunes. Like many large Hindu landowners in what is now Bangladesh, Radhika's family left for India around the time of Partition. They returned after Bangladesh's war of Independence in 1971, but they are unlikely to have fully recovered their former wealth from the period prior to their migration.

15. Shrimp exports from Bangladesh have declined over the past two decades both in value and quantity. In 2017–2018, export earnings fell for a fourth consecutive year to $408 million, accounting for the lowest quantity of exports since 2001–2002 (Parvez 2018). The variety of saltwater shrimp produced commercially in Bangladesh is giant tiger shrimp, or *Penaeus monodon*, which is known locally as *bagda*. This variety grows larger than the whiteleg shrimp, or *Litopenaeus vannamei*, which is the most commonly cultivated variety of tropical shrimp globally. *Vannamei* grows well in intensive conditions, which entail even more significant investments, thus allowing for greater yields per hectare, as is more common in these other large shrimp-producing countries. *Bagda* grows "extensively," meaning that it takes up more space, with lower yields but results in shrimp

that are larger in size. This means that they are considered higher quality and higher cost; as a result, economic downturns often cause international buyers to shift to the lower-cost variety. The large *bagda* shrimp, one WorldFish employee told me, using an Australian colloquialism, "is the kind you want to put on the barbie."

16. Throughout rural Bangladesh, it is common for people to be forced to migrate away from their homes when they are unable to repay microcredit loans. See Paprocki (2016) for more on this process of dispossession.

17. See also Paprocki and Cons (2014) on the relationship between shrimp aquaculture and dispossession and depeasantization.

18. Their failure to seek treatment could also have to do with the miserable conditions at Paikgachha's government hospital and its limited capacity. According to the Bangladesh Bureau of Statistics, the official number beds in this hospital is 60 (BBS 2013, 83), serving a total *upazila* population of about 250,000. There is also a government community clinic where women can seek treatment, but women in Kolanihat said it was very far for them to travel, with transport there costing them 20 taka ($.25), which many felt was out of reach.

19. Thus identified through a snowball sample from initial contacts in Kolanihat, I regard these interviews as an extension of my research in Kolanihat, not as a representative sample of migrants in Kolkata.

20. At this time, the national minimum wage for garment workers in Bangladesh was 5,300 taka (or about $62), though it was raised to 8,000 taka in December 2018. Men are typically paid more than women, who make up the vast majority of the workforce.

21. As Chatterjee and Hashmi have explained, this identification of the poorest members of the rural society with the peasantry of Bengal at large was fundamental to nationalist politics in anticolonial struggles as well as in the formation of an independent Pakistani state (Chatterjee 1994; Hashmi 1992). For Hashmi, this historical identity formation affirms Shanin's theses on the peasant class as a political force (Shanin 1966). Hashmi explains that this unity of the collective identity of the peasantry was particularly strong in Khulna historically because the rent-receiving *zamindars* and *talukdars* "had to be 'benevolent' landlords, as they needed the support of their tenants in the construction of embankments or *bundhs* to prevent saline water from entering the fields" (Hashmi 1992, 39). Chatterjee explains, however, that this unity was extended only to the Muslim rent-receivers, while Hindu landlords were not considered part of the peasant community (Chatterjee 1994, 11). Chatterjee argues that this aspect of peasant consciousness, grounded in "an entire set of beliefs about nature and about men in the collective and active mind of a peasantry," is fundamentally religious, rather than being based on a consciousness of class differences within the peasantry. He traces the communalism in Bengal that in part led to the formation of independent Indian and Pakistani states to this particular form of peasant consciousness in the region.

6. "WE HAVE COME THIS FAR—WE CANNOT RETREAT"

1. Exact dates for the end of shrimp cultivation are difficult to pinpoint both because residents' recollections vary and because the transition back to rice was slow and did not take place all at once. Given the ongoing conflicts over water management, it would be fair to say that the transition is ongoing even now. For the most part, however, farmers in Tilokpur shifted back to rice from shrimp sometime around 2009–2010, with a number of stops and starts both before and after.

2. This insight into the gendered division of agrarian labor draws on longstanding interventions in the feminist agrarian studies literature (Carney and Watts 1990; Hart 1992; Razavi 2009).

3. A good rice yield during this season is considered to be approximately 220–270 kg per acre, depending on the variety of seed used.

4. For reasons explored in chapters 3 and 4, most assumptions about the possibility and process of return to agriculture in communities like Tilokpur are entirely speculative and subject to the politics of uncertainty.

5. Clarke et al. (2015) found that when rice is irrigated during the dry season with water with a salinity concentration up to 4 parts per thousand (ppt), the monsoon rains are sufficient to flush the salinity from the soil; concentrations above 5 ppt cause an accumulation of salt in the soil and are therefore untenable for dry season rice irrigation. These findings are, however, contingent on the assumption that adequate drainage and water management systems are in place, which is generally not the case with shrimp ghers.

6. A rickshaw-van is a bicycle that pulls an attached flatbed, which is used for hauling produce and other cargo.

7. There are considered to be three primary growing seasons in Tilokpur; however, given the variable schedules of different vegetable harvests, a rickshaw might be needed to transport harvests more frequently depending on the diversity of planted crops and their growing seasons in a given year. Rickshaw-van drivers are also occasionally hired by village residents for personal transport or other needs.

8. The headmaster's agitation was also likely due in part to having his authority openly challenged in front of a foreigner (myself). It is also certainly possible that my presence in the room played a part in the success of their request to have the fees adjusted and that in other circumstances their demands may have required additional or different forms of pressure.

9. While a full examination of the field of Climate Smart Agriculture (CSA) is beyond the scope of this discussion, related analyses that explore the agrarian political economy of CSA (and its foundations in the Green Revolution orthodoxy of increased productivity) include Bezner Kerr 2012; Borras and Franco 2018; Newell and Taylor 2018; and Taylor 2018.

10. Some hybrid seed varieties originally developed by the Bangladesh Rice Research Institute (BRRI) have been reindigenized (what ecologists would call "naturalized") in the sense that farmers have saved and replanted them for many years. Naturalized hybrid seeds will not have the same traits as their predecessors, and these subsequent generations gradually adapt to their local environment, evolving away from the original variety. Thus, even as they call these seeds by the names given to them by BRRI (e.g., "BRRI 23"), farmers also often refer to these as *deshi* (meaning indigenous) as opposed to *hybrid*.

11. The Consultation Workshop is another term for the Stakeholder Workshop mentioned in the Methodological Appendix. Such workshops are considered obligatory components of the project conception process, although they are usually held after significant planning work has been carried out and funding commitments have been made for particular project components.

12. This includes a wide variety of hybrid seeds introduced through small demonstration plots. Of the plants produced by these seeds, one landless group member said to me dismissively, "They're nice to look at, but not nice to eat."

13. Relatedly, Blythe et al. have identified risks associated with discourses of "transformation" as apolitical or inevitable (2018).

14. BRAC (formerly the Bangladesh Rural Advancement Committee) is the largest NGO in the world.

15. See also Elliott (2017) on emerging moral economies of climate change, Rademacher (2018) on moral ecologies of sustainability discourse among environmental architects, and Angelo (2019) on social imaginaries of "greening."

CONCLUSION

1. See Alex Loftus's work on Gramscian political ecologies for more on political ecologies of possibility (Loftus 2009a, 2009b).

2. This discussion of enframing draws on Mitchell (1988, 2002).

3. See also West's (2020) call for understanding landscapes as palimpsests to both center their histories and serve as possibilities for the future.

4. See Tenzing (2020) for a review of literature on climate change adaptation and social protection.

5. This is not intended to suggest that the impacts of climate change are not being felt today. Indeed, today some impacts of climate change are unavoidable. Yet, even the effects of "carbon lock-in" are the result of the inertia of capitalist economic systems that resist reform, as opposed to a physical imperative that defies any possible intervention (Unruh 2000).

Bibliography

Abrams, Philip. 1988. "Notes on the Difficulty of Studying the State." *Journal of Historical Sociology* 1 (1): 58–89.

Abu-Lughod, Lila. 2000. "Locating Ethnography." *Ethnography* 1 (2): 261–267.

Adams, Vincanne. 2013. *Markets of Sorrow, Labors of Faith*. Durham, NC: Duke University Press.

ADB. 2012. "Addressing Climate Change and Migration in Asia and the Pacific." Asian Development Bank (Manila, Philippines).

Adnan, Shapan. 1991. *Floods, People and the Environment*. Dhaka: Research & Advisory Services.

———. 1993. "Shrimp-Culture Projects in Coastal Polders of Bangladesh: Policy Issues Concerning Socio-Economic and Environmental Consequences." Shomabesh Institute (Dhaka).

———. 1999. "Agrarian Structure and Agricultural Growth Trends in Bangladesh: The Political Economy of Technological Change and Policy Interventions." In *Sonar Bangla? Agricultural Growth and Agrarian Change in West Bengal and Bangladesh*, edited by Ben Rogaly, Barbara Harriss-White, and Sugata Bose, 177–228. New Delhi: Sage Publications.

———. 2009. "Intellectual Critiques, People's Resistance and Inter-Riparian Contestations: Constraints to the Power of the State Regarding Flood Control and Water Management in the Ganges-Brahmaputra-Meghna Delta of Bangladesh." In *Water, Sovereignty and Borders: Fresh and Salt in Asia and Oceania*, edited by Devleena Ghosh, Heather Goodall. and Stephanie Hemelryk Donald, 104–124. London: Routledge.

———. 2013. "Land Grabs and Primitive Accumulation in Deltaic Bangladesh: Interactions between Neoliberal Globalization, State Interventions, Power Relations and Peasant Resistance." *Journal of Peasant Studies* 40 (1): 87–128.

Adnan, Shapan, Alison Barrett, S. M. Nurul Alam, and Angelika Brustinow. 1992. "People's Participation, NGOs and the Flood Action Plan: An Independent Review." Research & Advisory Services (Dhaka).

Advisory Group on Development of Deltaic Areas. 1966. "Appraisal of Some Aspects of the Coastal Embankment Project of East Pakistan." United Nations Economic Commission for Asia and the Far East, February 1966.

Afroz, Sharmin, Rob Cramb, and Clemens Grünbühel. 2016. "Collective Management of Water Resources in Coastal Bangladesh: Formal and Substantive Approaches." *Human Ecology* 44: 17–31.

———. 2017. "Exclusion and Counter-Exclusion: The Struggle over Shrimp Farming in a Coastal Village in Bangladesh." *Development and Change* 48 (4): 692–720.

Afroz, Tanzim, and Shawkat Alam. 2013. "Sustainable Shrimp Farming in Bangladesh: A Quest for an Integrated Coastal Zone Management." *Ocean & Coastal Management* 71: 275–283.

Agency for International Development. 1967. "Pakistan: Consulting Services for East Pakistan Water and Power Development Authority." Department of State Agency for International Development (Washington, DC).

Ahmed, Asib. 2011. "Some of the Major Environmental Problems Relating to Land Use Changes in the Coastal Areas of Bangladesh: A Review." *Journal of Geography and Regional Planning* 4 (1): 1–8.

Ahmed, Farid Hasan. 2017. "How We Can Prepare Ourselves for Cyclones." *Daily Star* (Bangladesh), May 31, 2017.

Ahuja, Neel. 2016. "Race, Human Security, and the Climate Refugee." *English Language Notes* 54 (2): 25–32.

———. 2017. "Posthuman New York: Ground Zero of the Anthropocene." In *Animalities: Literary and Cultural Studies beyond the Human*, edited by Michael Lundblad, 43–59. Edinburgh: Edinburgh University Press.

Akhter, Majed. 2015. "The Hydropolitical Cold War: The Indus Waters Treaty and State Formation in Pakistan." *Political Geography* 46: 65–75.

Akram-Lodhi, A. Haroon, and Cristóbal Kay, eds. 2009. *Peasants and Globalization: Political Economy, Rural Transformation and the Agrarian Question*. London: Routledge.

Akram-Lodhi, A. Haroon, Saturnino M. Borras, Jr., and Cristóbal Kay, eds. 2006. *Land, Poverty and Livelihoods in an Era of Globalization: Perspectives from Developing and Transition Countries*. London: Routledge.

Akter, Sonia, and Bishwajit Mallick. 2013. "The Poverty-Vulnerability-Resilience Nexus: Evidence from Bangladesh." *Ecological Economics* 96: 114–124.

Alauddin, Mohammad, and MA Hamid. 1999. "Coastal aquaculture in South Asia: Experiences and lessons." In *Development, governance and environment in South Asia: A focus on Bangladesh*, edited by Mohammad Alauddin and Samiul Hasan, 289–299. London: Palgrave MacMillan.

Alexander, Claire, Joya Chatterji, and Annu Jalais. 2016. *The Bengal Diaspora: Rethinking Muslim Migration*. London: Routledge.

Ali, Tariq Omar. 2019. "Technologies of Peasant Production and Reproduction: The Postcolonial State and Cold War Empire in Comilla, East Pakistan, 1960–1970." *South Asia Journal* 41 (3): 435–451.

Allison, M. A., S. R. Khan, S. L. Goodbred Jr., and S. A. Kuehl. 2003. "Stratigraphic Evolution of the Late Holocene Ganges-Brahmaputra Lower Delta Plain." *Sedimentary Geology* 155: 317–342.

Amery, C. F. 1876. "On Forest Rights in India." In *Report on the Proceedings of the Forest Conference Held at Simla*, edited by D. Brandis and A. Smythies, 27–30. Calcutta: Office of the Superintendent of Government Printing.

Angelo, Hillary. 2019. "The Greening Imaginary: Urbanized Nature in Germany's Ruhr Region." *Theory and Society* 48 (5): 645–669.

Anisuzzaman, M., Badaruddin Ahmed, A. K. M. Serajul Islam, and Sajjad Hussain. 1986. *Comilla Models of Rural Development: A Quarter Century of Experience*. Comilla: BARD.

Anonymous. 1859. "The Gangetic Delta." *Calcutta Review* 63: 1–20.

Ariza-Montobbio, Pere, Sharachchandra Lele, Giorgios Kallis, and Joan Martinez-Alier. 2010. "The Political Ecology of Jatropha Plantations for Biodiesel in Tamil Nadu, India." *Journal of Peasant Studies* 37 (4): 875–897.

Arora-Jonsson, Seema, Lisa Westholm, Beatus John Temu, and Andrea Petitt. 2016. "Carbon and Cash in Climate Assemblages: The Making of a New Global Citizenship." *Antipode* 48 (1): 74–96.

Asian Development Bank. 2007. "Project Performance Evaluation Report in Bangladesh: Khulna-Jessore Drainage Rehabilitation Project." Asian Development Bank, Operations Evaluation Department, December 2007.

Auerbach, L. W., S. L. Goodbred Jr., D. R. Mondal, C. A. Wilson, K. R. Ahmed, K. Roy, M. S. Steckler, C. Small, J. M. Gilligan, and B. A. Ackerly. 2015a. "Flood Risk of Natural and Embanked Landscapes on the Ganges-Brahmaputra Tidal Delta Plain." *Nature Climate Change* 5 (2): 153–157.
———. 2015b. "Reply to 'Tidal River Management in Bangladesh.'" *Nature Climate Change* 5 (6): 492–493.
Bagchi, Suvojit. 2014. "On the Dark Side of Kolkata's Satellite Towns." *The Hindu*, June 21, 2014.
Bahadur, Aditya, and Thomas Tanner. 2014. "Transformational Resilience Thinking: Putting People, Power and Politics at the Heart of Urban Climate Resilience." *Environment & Urbanization* 26 (1): 200–214.
Baka, Jennifer. 2013. "The Political Construction of Wasteland: Governmentality, Land Acquisition and Social Inequality in South India." *Development and Change* 44 (2): 409–428.
———. 2017. "Making Space for Energy: Wasteland Development, Enclosures, and Energy Dispossessions." *Antipode* 49 (4): 977–996.
Ballu, Valérie, Marie-Noëlle Bouin, Patricia Siméoni, Wayne C. Crawford, Stephane Clamant, Jean-Michel Boré, Tony Kanas, and Bernard Pelletier. 2011. "Comparing the Role of Absolute Sea-Level Rise and Vertical Tectonic Motions in Coastal Flooding, Torres Islands (Vanuatu)." *PNAS* 108 (32): 13019–13022.
Bandyopadhyay, A. K. 1987. "A Note on Effect of Reclamation through Marginal Embankments on Coastal Eco-system of Sunderban." *Journal of the Indian Society of Coastal Agricultural Research* 5 (1): 39–41.
Bari, K. G. M. Latiful. 1978. *Bangladesh District Gazeteers: Khulna*. Dacca: Bangladesh Government Press.
Barnes, Jessica, and Michael R. Dove, eds. 2015. *Climate Cultures: Anthropological Perspectives on Climate Change*. New Haven, CT: Yale University Press.
Barraclough, Solon, and Andrea Finger-Stich. 1996. "Some Ecological and Social Implications of Commercial Shrimp Farming in Asia." *United Nations Research Institute for Social Development* 74: 71.
Barrow, Ian J. 2003. *Making History, Drawing Territory: British Mapping in India, c. 1756–1905*. Oxford: Oxford University Press.
Barua, Prabal, Md. Shah Nawaz Chowdhury, and Subrata Sarkar. 2010. "Climate Change and Its Risk Reduction by Mangrove Ecosystem of Bangladesh." *Bangladesh Research Publications Journal* 4 (3): 208–225.
Bassett, Thomas J., and Charles Fogelman. 2013. "Déjà Vu or Something New? The Adaptation Concept in the Climate Change Literature." *Geoforum* 48: 42–53.
Baviskar, Amita. 1995. *In the Belly of the River: Tribal Conflicts over Development in the Narmada Valley*. Delhi: Oxford University Press.
BBS. 2012. "Community Report: Khulna Zila." Bangladesh Bureau of Statistics, Statistics and Informatics Division, Ministry of Planning.
———. 2013. "District Statistics 2011: Khulna." Dhaka: Bangladesh Bureau of Statistics.
———. 2016. *Statistical Pocketbook Bangladesh 2015*. Dhaka: Bangladesh Bureau of Statistics.
Beck, Silke, and Martin Mahony. 2018. "The Politics of Anticipation: the IPCC and the Negative Emissions Technologies Experience." *Global Sustainability* 1: e8.
Belton, Ben. 2016. "Shrimp, Prawn and the Political Economy of Social Wellbeing in Rural Bangladesh." *Journal of Rural Studies* 45: 230–242.
Béné, Christophe. 2005. "The Good, the Bad and the Ugly: Discourse, Policy Controversies and the Role of Science in the Politics of Shrimp Farming Development." *Development Policy Review* 23 (5): 585–614.

Bengal Government. 1898. *Appendices to the Final Resolution of the Government of Bengal Upon the Famine of 1896 and 1897.* Vol. 3. Calcutta: Bengal Secretariat Press.

Bernstein, Henry. 2006. "Is There an Agrarian Question in the 21st Century?" *Canadian Journal of Development Studies* 27 (4): 449–460.

Beveridge, H. 1876a. *The District of Bákarganj: Its History and Statistics.* London: Trübner & Co.

———. 1876b. "Were the Sundarbans Inhabited in Ancient Times?" *Journal of the Asiatic Society of Bengal* (1): 71–76.

Bezner Kerr, Rachel. 2012. "Lessons from the Old Green Revolution for the New: Social, Environmental and Nutritional Issues for Agricultural Change in Africa." *Progress in Development Studies* 12 (2–3): 213–229.

Bhattacharyya, Debjani. 2018a. "Discipline and Drain: Settling the Moving Bengal Delta." *Global Environment* 11: 236–257.

———. 2018b. *Empire and Ecology in the Bengal Delta: The Making of Calcutta.* Cambridge: Cambridge University Press.

Bhattacharyya, Jnanabrata. 1990. "Uses, Values, and Use Values of the Sundarbans." *Agriculture and Human Values* 7 (2): 34–39.

Bjornberg, Anders. 2016. "Rohingya Territoriality in Myanmar and Bangladesh: Humanitarian Crisis and National Disordering." In *Myanmar's Mountain and Maritime Borderscapes: Local Practices, Boundary-Making and Figured Worlds,* edited by Su-Ann Oh, 146–167. Singapore: ISEAS Publishing.

Black, George. 2013. "Your Clothes Were Made by a Bangladeshi Climate Refugee." *Mother Jones,* July 30, 2013.

Black, Megan. 2018. *The Global Interior: Mineral Frontiers and American Power.* Cambridge, MA: Harvard University Press.

Blaikie, Piers. 1985. *The Political Economy of Soil Erosion in Developing Countries.* London: Longman Development Series.

Blair, Harry. 1978. "Rural Development, Class Structure and Bureaucracy in Bangladesh." *World Development* 6 (1): 65–82.

Blood, Archer, Richard Nixon, and Henry Kissinger. 2013. "A Telegram and a Phone Call." In *The Bangladesh Reader: History, Culture, Politics,* edited by Meghna Guhathakurta and Willem van Schendel, 241–243. Durham: Duke University Press.

Blythe, Jessica, Jennifer Silver, Louisa Evans, Derek Armitage, Nathan J. Bennett, Michele-Lee Moore, Tiffany H. Morrison, and Katrina Brown. 2018. "The Dark Side of Transformation: Latent Risks in Contemporary Sustainability Discourse." *Antipode* 50 (5): 1206–1223.

Bond, Patrick. 2012. "Emissions Trading, New Enclosures and Eco-Social Contestation." *Antipode* 44 (3): 684–701.

Borras, Saturnino M., Jr., and Jennifer C. Franco. 2012. "Global Land Grabbing and Trajectories of Agrarian Change: A Preliminary Analysis." *Journal of Agrarian Change* 12 (1): 34–59.

———. 2018. "The Challenge of Locating Land-Based Climate Change Mitigation and Adaptation Politics within a Social Justice Perspective: Towards an Idea of Agrarian Climate Justice." *Third World Quarterly* 39 (7): 1308–1325.

Bose, Sugata. 1986. *Agrarian Bengal: Economy, Social Structure and Politics, 1919–1947.* Cambridge: Cambridge University Press.

———. 1993. *Peasant Labour and Colonial Capital: Rural Bengal since 1770.* Cambridge: Cambridge University Press.

———. 1999. "Agricultural Growth and Agrarian Structure in Bengal: A Historical Overview." In *Sonar Bangla? Agricultural Growth and Agrarian Change in West Bengal*

and Bangladesh, edited by Ben Rogaly, Barbara Harriss-White, and Sugata Bose, 41–59. New Delhi: Sage Publications.
Bose, Swadesh R. 1972. "East-West Contrast in Pakistan's Agricultural Development." In *Growth and Inequality in Pakistan*, edited by Keith Griffin and Azizur Rahman Khan, 69–93. London: Palgrave Macmillan.
Boyce, James. 1987. *Agrarian Impasse in Bengal: Institutional Constraints to Technological Change*. Oxford: Oxford University Press.
———. 1990. "Birth of a Megaproject: Political Economy of Flood Control in Bangladesh." *Environmental Management* 14 (4): 419–428.
BRAC. September 2013. "Alternative Livelihood Options for Climate Change Vulnerable Women Group." BRAC.
Bradnock, Robert W., and Patricia L. Saunders. 2000. "Sea-Level Rise, Subsidence and Submergence: The Political Ecology of Environmental Change in the Bengal Delta." In *Political Ecology: Science, Myth and Power*, edited by Philip Anthony Stott and Sian Sullivan, 66–90. Oxford: Oxford University Press.
Brammer, Hugh. 2009. "Climate Refugees: A Rejoinder." *Economic and Political Weekly* 44 (29): 87.
———. 2012. *The Physical Geography of Bangladesh*. Dhaka: University Press Limited.
———. 2014a. "Bangladesh's Dynamic Coastal Regions and Sea-level Rise." *Climate Risk Management* 1: 51–62.
———. 2014b. *Climate Change, Sea-Level Rise and Development in Bangladesh*. Dhaka: University Press Limited.
Brandis, D., and A. Smythies, eds. 1876. *Report of the Proceedings of the Forest Conference Held at Simla, October 1875*. Calcutta: Office of the Superintendent of Government Printing.
Brown, S., and R. J. Nicholls. 2015. "Subsidence and Human Influences in Mega-deltas: The Case of the Ganges-Brahmaputra-Meghna." *Science of the Total Environment* 527–528: 362–374.
Brown, Sally, Robert J. Nicholls, Susan Hanson, Geoff Brundrit, John A. Dearing, Mark E. Dickson, Shari L. Gallop, Shu Gao, Ivan D. Haigh, Jochen Hinkel, and José A. Jiménez. 2014. "Shifting Perspectives on Coastal Impacts and Adaptation." *Nature Climate Change* 4 (9): 752–755.
Buckely, Robert Burton. 1906. "The Navigable Waterways of India." *Journal of the Society of Arts* 54 (2780): 417–458.
Buckland, C. E. 1901. *Bengal under the Lieutenant-Governors*. Vol. 2. Calcutta: S. K. Lahiri & Co.
Bull, John. 1823. *The Calcutta Annual Register for the Year 1821*. Calcutta: Government Gazette Press.
Byron, Rezaul Karim. 2015. "Budget Support Merits Stern Reforms: WB." *Daily Star* (Bangladesh), April 19, 2015. Business section.
Camargo, Alejandro. 2017. "Land Born of Water: Property, Stasis, and Motion in the Floodplains of Northern Colombia." *Geoforum*, November 20, 2017. https://doi.org/10.1016/j.geoforum.2017.11.006.
Carney, J., and Michael Watts. 1990. "Manufacturing Dissent: Work, Gender and the Politics of Meaning in a Peasant Society." *Africa* 60 (2): 207–241.
Castree, Noel. 2015. "Geographers and the Discourse of an Earth Transformed: Influencing the Intellectual Weather or Changing the Intellectual Climate?" *Geographical Research* 53 (3): 244–254.
Chakraborti, Suman. 2013. "New Town Inches Closer to Solar City." *Times of India*, October 2, 2013.

———. 2014. "West Bengal Seeks 'Smart Green City' Tag for New Town." *Times of India*, July 6, 2014.
Chapman, Graham. 2007. *The Geopolitics of South Asia*. Farnham, UK: Ashgate.
Chatterjee, Partha. 1994. "Agrarian Relations and Communalism in Bengal, 1926–1935." In *Subaltern Studies I: Writings on South Asian History and Society*, edited by Ranajit Guha, 9–38. Oxford: Oxford University Press.
Chattopadhyaya, Haraprasad. 1999. *The Mystery of the Sundarbans*. Calcutta: A Mukherjee & Co.
Chiu, Soyee, and Christopher Small. 2016. "Observations of Cyclone-Induced Storm Surge in Coastal Bangladesh." *Journal of Coastal Research* 32 (5): 1149–1161.
Choldin, Harvey M. 1969. "The Development Project as Natural Experiment: The Comilla, Pakistan, Projects." *Economic Development and Cultural Change* 17 (4): 483–500.
Choudhury, Moqbul A. 1969. *The Comilla Family Planning Project: Seventh Progress Report, July 1968–June 1969*. Comilla: PARD.
Claeys, Gregory. 2017. *Dystopia: A Natural History*. Oxford: Oxford University Press.
Clarke, D., S. Williams, M. Jahiruddin, K. Parks, and M. Salehin. 2015. "Projections of On-Farm Salinity in Coastal Bangladesh." *Environmental Science: Processes & Impacts* 17: 1127–1136.
Cohen, Daniel Aldana. 2016. "The Rationed City: The Politics of Water, Housing and Land Use in Drought-Parched São Paulo." *Public Culture* 28 (2): 261–289.
———. 2017. "The Other Low-Carbon Protagonist: Poor People's Movements and Climate Politics in São Paulo." In *The City Is the Factory: New Solidarities and Spatial Strategies in an Urban Age*, edited by Miriam Greenberg and Penny Lewis, 140–157. Ithaca, NY: Cornell University Press.
Cons, Jason. 2018. "Staging Climate Security: Resilience and Heterodystopia in the Bangladesh Borderlands." *Cultural Anthropology* 33 (2): 266–294.
Cons, Jason, and Kasia Paprocki. 2010. "Contested Credit Landscapes: Microcredit, Self-help and Self-Determination in Rural Bangladesh." *Third World Quarterly* 31 (4): 637–654.
Corbera, Esteve, Laura Calvet-Mir, Hannah Hughes, and Matthew Paterson. 2015. "Patterns of Authorship in the IPCC Working Group III Report." *Nature Climate Change* 6 (1): 94–99.
Curley, Andrew. 2018. "A Failed Green Future: Navajo Green Jobs and Energy 'Transition' in the Navajo Nation." *Geoforum* 88: 57–65.
Daily Star (Bangladesh). 2014a. "HSBC-Star Junior Climate Champions in CTG." September 13, 2014, City edition.
Daily Star (Bangladesh). 2014b. "Micro Loans Can Offset Effects of Climate Change: Analyst." November 16, 2014. Business section.
Daily Star (Bangladesh). 2014c. "Path to Middle-Income Nation: Organised Labour Market, Skills Building." May 6, 2014. Business section.
Darby, Stephen E., Frances E. Dunn, Robert J. Nicholls, Munsur Rahman, and Liam Riddy. 2015. "A First Look at the Influence of Anthropogenic Climate Change on the Future Delivery of Fluvial Sediment to the Ganges-Brahmaputra-Meghna Delta." *Environmental Science: Processes & Impacts* 17: 1587–1600.
Das, Gokul Chandra. 1996. "The Colonial Land-System in the 24 Parganas: Sundarbans, 1825–1904." *Bengal Past & Present* 115: 48–65.
Das, Tapan Kumar, and Ramkrishna Maiti. 2010. "Land Reclamation, Drainage Decay and Embankment Breaching—A Case Study along the Raimangal." *Indian Journal of Geography and Environment* 11: 19–29.
Dasgupta, Susmita, Mainul Huq, Zahirul Huq Khan, Manjur Murshed Zahid Ahmed, Nandan Mukherjee, Malik Fida Khan, and Kiran Pandey. 2014. "Cyclones in a Changing Cliamte: The Case of Bangladesh." *Climate and Development* 6 (2): 96–110.

Dasgupta, Susmita, Farhana Akhter Kamal, Zahirul Haque Khan, Sharifussaman Choudhury, and Ainun Nishat. 2014. "River Salinity and Climate Change: Evidence from Coastal Bangladesh." World Bank (Washington, DC).
Datta, Anjan Kumar. 1998. *Land and Labour Relations in South-West Bangladesh: Resources, Power and Conflict.* New York: St. Martin's Press.
———. 2006. "Who Benefits and at What Cost? Expanded Shrimp Culture in Bangladesh." In *Shrimp Farming and Industry: Sustainability, Trade and Livelihoods*, edited by A. Atiq Rahman, A. H. G. Quddus, Bob Pokrant, and Md. Liaquat Ali, 507–528. Dhaka: University Press Limited.
Davis, Peter, and Snigdha Ali. 2014. "Exploring Local Perceptions of Climate Change Impact and Adaptation in Rural Bangladesh." IFPRI discussion papers 1322, International Food Policy Research Institute, February 2014.
Deb, Apurba Krishna. 1998. "Fake Blue Revolution: Environmental and Socio-economic Impacts of Shrimp Culture in the Coastal Areas of Bangladesh." *Ocean & Coastal Management* 41: 63–88.
de Hoop, Evelien, and Saurabh Arora. 2017. "Material Meanings: 'Waste' as a Performative Category of Land in Colonial India." *Journal of Historical Geography* 55: 82–92.
Demeritt, David. 2001. "The Construction of Global Warming and the Politics of Science." *Annals of the Association of American Geographers* 91 (2): 307–337.
Derrington, Erin. 2015. "NELD Story by Erin Derrington." In *NELD: Non-Economic Loss and Damage.* http://climate-neld.com/neld-story-by-erin-derrington/.
de Sherbinin, Alex, Marcia C. de Castro, François Gemenne, Michael M. Cernea, Susan Beatriz Adamo, Philip Fearnside, Gary Krieger, Sarah Lahmani, Anthony Oliver-Smith, Alula Pankhurst, Thayer Scudder, Burton Singer, Yan Tan, Gregory E. Wannier, Philippe Boncour, Charles Ehrhart, Graeme Hugo, B. Pandey, and Guo Shi. 2011. "Preparing for Resettlement Associated with Climate Change." *Science* 334 (6055): 456–457.
Devine, Joseph. 2003. "The Paradox of Sustainability: Reflections on NGOs in Bangladesh." *Annals of the American Academy of Political and Social Science* 590 (Rethinking Sustainable Development): 227–242.
Dey, Ishita, Ranabir Samaddar, and Suhit K. Sen. 2013. *Beyond Kolkata: Rajarhat and the Dystopia of Urban Imagination.* Delhi: Routledge India.
Dey, Madan M., David J. Spielman, ABM Mahfuzul Haque, Md Saidur Rahman, and Rowena A Valmonte-Santos. 2012. *Change and Diversity in Smallholder Rice-Fish Systems: Recent Evidence from Bangladesh.* International Food Policy Research Institute.
Dhaka Tribune. 2014a. "Modi Seething at Bangladeshi Immigrants." April 29, 2014.
Dhaka Tribune. 2014b. "Bangladesh World's 2nd Most Pro–Free Market Country." November 1, 2014.
Dickens, Charles. 1875. "The Guns of Burrisaul." *All the Year Round* 14 (346): 378–380.
D'Souza, Rohan. 2015. "Mischievous Rivers and Evil Shoals: The English East India Company and the Colonial Resource Regime." In *The East India Company and the Natural World*, edited by Vinita Damodaran, Anna Winterbottom, and Alan Lester, 128–146. Basingstoke, Hampshire, UK: Palgrave Macmillan.
Duflo, Esther, Abhijit Banerjee, Rachel Glennerster, and Cynthia Kinnan. 2013. "The Miracle of Microfinance? Evidence from a Randomized Evaluation." NBER Working Paper No. 18950. US National Bureau of Economic Research (Cambridge, MA), May 2013.
Eales, H. L. 1892. *Census of 1891: Burma Report.* Vol. 1. Rangoon: Superintendent, Government Printing, Burma.
Eardley-Wilmot, Sainthill. 1910. *Forest Life and Sport in India.* London: Edward Arnold.

Eaton, Richard M. 1990. "Human Settlement and Colonization in the Sundarbans, 1200–1750." *Agriculture and Human Values* 7 (2): 12–31.

———. 1993. *The Rise of Islam and the Bengal Frontier, 1204–1760*. Berkeley: University of California Press.

Edelman, Marc. 2013. "Messy Hectares: Questions about the Epistemology of Land Grabbing Data." *Journal of Peasant Studies* 40 (3): 485–501.

Elliott, Rebecca. 2017. "Who Pays for the Next Wave? The American Welfare State and Responsibility for Flood Risk." *Politics & Society* 45 (3): 415–440.

———. 2019. "'Scarier Than Another Storm': Values at Risk in the Mapping and Insuring of US Floodplains." *British Journal of Sociology* 70 (3): 1067–1090.

Eriksen, Siri H., Andrea J. Nightingale, and Hallie Eakin. 2015. "Reframing Adaptation: The Political Nature of Climate Change Adaptation." *Global Environmental Change* 35: 523–533.

Faaland, Just, and J. R. Parkinson. 1976. *Bangladesh: The Test Case of Development*. London: C Hurst and Company.

FAO. March 2015. "Mapping Exercise on Water-Logging in South West of Bangladesh: Draft for Consultation." Food and Agriculture Organization of the United Nations (Dhaka).

Farbotko, Carol. 2010a. "'The Global Warming Clock Is Ticking so See These Places While You Can': Voyeuristic Tourism and Model Environmental Citizens on Tuvalu's Disappearing Islands." *Singapore Journal of Tropical Geography* 31 (2): 224–238.

———. 2010b. "Wishful Sinking: Disappearing Islands, Climate Refugees and Cosmopolitan Experimentation." *Asia Pacific Viewpoint* 51 (1): 47–60.

Fawcus, L. R. 1927. *Final Report on the Khulna Settlement 1920–1926*. Calcutta: Bengal Secretariat Book Depot.

Federici, Silvia. 2004. *Caliban and the Witch: Women, the Body and Primitive Accumulation*. London: Autonomedia.

Feldman, Shelley. 1997. "NGOs and Civil Society: (Un)stated Contradictions." *Annals of the American Academy of Political Science* 554: 46–65.

———. 1999. "Feminist Interruptions: The Silence of East Bengal in the Story of Partition." *Interventions: International Journal of Postcolonial Studies* 1 (2): 167–182.

———. 2003. "Paradoxes of Institutionalization: The Depoliticisation of Bangladeshi NGOs." *Development in Practice* 13 (1): 23.

Feldman, Shelley, and Florence McCarthy. 1984. "Constraints Challenging the Cooperative Strategy in Bangladesh." *South Asia Bulletin* 4 (2): 11–21.

Felli, Romain, and Noel Castree. 2012. "Neoliberalising Adaptation to Environmental Change: Foresight or Foreclosure?" *Environment and Planning A* 44 (1): 1–4.

Ferguson, James. 1994. *The Anti-Politics Machine: Development, Depoliticization, and Bureaucratic Power in Lesotho*. Cambridge: Cambridge University Press.

Field, C. D. 1885. *Landholding and the Relation of Landlord and Tenant, in Various Countries*. Calcutta: Thacker, Spink and Co.

Ford, Allison, and Kari Marie Norgaard. 2020. "Whose Everyday Climate Cultures? Environmental Subjectivities and Invisibility in Climate Change Discourse." *Climatic Change* 163: 43–62.

Forsyth, Tim. 2013. "Community-Based Adaptation: A Review of Past and Future Challenges." *WIREs Climate Change* 4 (5): 439–446.

———. 2014. "Climate Justice Is Not Just Ice." *Geoforum* 54: 230–232.

Foucault, Michel. 1980. *Power/Knowledge: Selected Interviews and Other Writings 1972–1977*, edited by Colin Gordon. New York: Pantheon Books.

Friedmann, Harriet, and Philip McMichael. 1989. "Agriculture and the State System: The Rise and Decline of National Agricultures, 1870 to the Present." *Sociologia Ruralis* 29 (2): 93–117.

Gasper, Des. 2000. "Evaluating the 'Logical Framework Approach' towards Learning-Oriented Development Evaluation." *Public Administration and Development* 20 (1): 17–28.
Gastrell, J. E. 1868. *Statistical Report of the Districts of Jessore, Fureedpore and Backergunge*. Calcutta: Office of Superintendent of Government Printing.
Gidwani, Vinay. 1992. "'Waste' and the Permanent Settlement in Bengal." *Economic and Political Weekly* 27 (4): PE-39–PE-46.
———. 2012. "Waste/Value." In *The Wiley-Blackwell Companion to Economic Geography*, edited by Trevor Barnes, Jamie Peck, and Eric Sheppard, 275–288. Malden, MA: Blackwell Publishing.
Gidwani, Vinay, and Rajyashree N. Reddy. 2011. "The Aferlives of 'Waste': Notes from India for a Minor History of Capitalist Surplus." *Antipode* 43 (5): 1625–1658.
Gilchrist, J. B. 1825. *The General East India Guide and Vade Mecum*. London: Kingsbury, Parbury, & Allen.
Gilmartin, David. 2003. "Water and Waste: Nature, Productivity and Colonialism in the Indus Basin." *Economic and Political Weekly* 38 (48): 5057–5065.
Ginn, Franklin. 2015. "When Horses Won't Eat: Apocalypse and the Anthropocene." *Annals of the Association of American Geographers* 105 (2): 1–9.
Glennon, Robert. 2017. "The Unfolding Tragedy of Climate Change in Bangladesh." Guest blog. *Scientific American*, April 21, 2017. https://blogs.scientificamerican.com/guest-blog/the-unfolding-tragedy-of-climate-change-in-bangladesh/.
Goh, Kian. 2020. "Flows in Formation: The Global-Urban Networks of Climate Change Adaptation." *Urban Studies* 57 (11): 2222–2240.
Goldman, Michael. 2005. *Imperial Nature: The World Bank and Struggles for Social Justice in the Age of Globalization*. New Haven, CT: Yale University Press.
Goldstein, Jesse. 2012. "*Terra Economica*: Waste and the Production of Enclosed Nature." *Antipode* 45 (2): 357–375.
Gotham, Kevin Fox. 2012. "Disaster, Inc.: Privatization and Post-Katrina Rebuilding in New Orleans." *Perspectives on Politics* 10 (3): 633–646.
Gramsci, Antonio. 1971. *Selections from the Prison Notebooks*. Moscow: International Publishers.
Greenough, Paul. 1998. "Hunter's Drowned Land: An Environmental Fantasy of the Victorian Sunderbans." In *Nature and the Orient: The Environmental History of South and Southeast Asia*, edited by Richard H. Grove, Vinita Damodaran, and Satpal Sangwan, 237–272. Delhi: Oxford University Press.
Guha, Ramachandra. 1990. "An Early Environmental Debate: The Making of the 1878 Forest Act." *Indian Economic and Social History Review* 27 (1): 65–84.
Guha, Ranajit. 1982. *A Rule of Property for Bengal: An Essay on the Idea of Permanent Settlement*. Andhra Pradesh: Orient Blackswan.
Guhathakurta, Meghna. 2008. "Globalization, Class and Gender Relations: The Shrimp Industry in Southwestern Bangladesh." *Development* 51 (2): 212–219.
———. 2011. "The Gendered Nature of Migration in Southwestern Bangladesh: Lessons for a Climate Change Policy." Paper presented at the conference Rethinking Migration: Climate, Resource Conflicts and Migration in Europe, Berlin, Germany, October 13–14.
Günel, Gökçe. 2016. "What Is Carbon Dioxide? When Is Carbon Dioxide?" *PoLAR* 39 (1): 33–45.
Gupta, Akhil. 1998. *Postcolonial Developments: Agriculture in the Making of Modern India*. Durham, NC: Duke University Press.
Gupta, Pramodranjan Das. 1935. *Supplement to the Final Report on the Survey and Settlement Operations in the District of Khulna for the Period 1926–1927*. Alipore, Bengal: Bengal Government Press.

Haalboom, Bethany, and David C. Natcher. 2012. "The Power and Peril of 'Vulnerability': Approaching Community Labels with Caution in Climate Change Research." *ARCTIC* 65 (3): 319–327.

Hall, Stuart. 1980. "Race, Articulation, and Societies Structured in Dominance." In *Sociological Theories: Race and Colonialism*, edited by United Nations Educational. Scientific, and Cultural Organisation, 305–345. Paris: UNESCO.

———. 1985. "Signification, Representation, Ideology: Althusser and the Post-Structuralist Debates." *Critical Studies in Mass Communication* 2 (2): 91–114.

Hamilton, Walter. 1820. *A Geographical, Statistical, and Historical Description of Hindostan, and the Adjacent Countries*. Vol. 1. London: John Murray.

———. 1828. *The East-India Gazetteer*. Vol. 1. London: Parbury, Allen, and Co.

Hanebuth, Till J. J., Hermann R. Kudrass, Jörg Linstädter, Badrul Islam, and Anja M. Zander. 2013. "Rapid Coastal Subsidence in the Central Ganges-Brahmaputra Delta (Bangladesh) since the 17th Century Deduced from Submerged Salt-Producing Kilns." *Geology* 41 (9): 987–990.

Haq, Mazharul, Nurul Islam, Rehman Sobhan, M. Akhlaqur Rehman, A. K. M. Ghulam Rabbani, and Anisur Rehman. 1976. "Growth and Causes of Regional Disparity in Pakistan." *Pakistan Economic and Social Review* 14 (1/4): 265–290.

Haque, Anika Nasra, David Dodman, and Md. Mohataz Hossain. 2014. "Individual, Communal and Institutional Responses to Climate Change by Low-income Households in Khulna, Bangladesh." *Environment & Urbanization* 26 (1): 112–129.

Haque, Moinul. 2014. "Bangladesh Shrimp Exports to US Market Plummet." *New Age Bangladesh*, September 21, 2014.

Haque, S. A. 2006. "Salinity Problems and Crop Production in Coastal Regions of Bangladesh." *Pakistan Journal of Botany* 38 (5): 1359–1365.

Haque, Wahidul. 1977. "Toward a Theory of Rural Development." *Development Dialogue* 2: 89–103.

Hardee, Karen, Sandor Balogh, and Michele T. Villinski. 1997. "Three Countries' Experience with Norplant Introduction." *Health Policy and Planning* 12 (3): 199–213.

Harms, Erik, and Ian G. Baird. 2014. "Wastelands, Degraded Lands and Forests, and the Class(ification) Struggle: Three Critical Perspectives from Mainland Southeast Asia." *Singapore Journal of Tropical Geography* 35: 289–294.

Harris, Gardiner. 2014. "As Seas Rise, Millions Cling to Borrowed Time and Dying Land." *New York Times*, March 29, 2014.

Harrison, Henry Leland. 1875. *The Bengal Embankment Manual*. Calcutta: Bengal Secretariat Press.

Hart, Gillian. 1992. "Household Production Reconsidered: Gender, Labor Conflict, and Technological Change in Malaysia's Muda Region." *World Development* 20 (6): 809–823.

———. 2001. "Development Critiques in the 1990s: *Culs de Sac* and Promising Paths." *Progress in Human Geography* 24 (4): 649–658.

———. 2002a. *Disabling Globalization: Places of Power in Post-Apartheid South Africa*. Berkeley: University of California Press.

———. 2002b. "Geography and Development: Development/s beyond Neoliberalism? Power, Culture, Political Economy." *Progress in Human Geography* 26 (6): 812–822.

———. 2018. "Relational Comparison Revisited: Marxist Postcolonial Geographies in Practice." *Progress in Human Geography* 42 (3): 371–394.

Hartmann, Betsy. 1995. *Reproductive Rights and Wrongs: The Global Politics of Population Control and Contraceptive Choice*. Cambridge, MA: South End Press.

———. 2010. "Rethinking Climate Refugees and Climate Conflict: Rhetoric, Reality and the Politics of Policy Discourse." *Journal of International Development* 22 (2): 233–246.

Hartmann, Betsy, and James K. Boyce. 1983. *A Quiet Violence: View from a Bangladesh Village*. Dhaka: University Press Limited.
Hashmi, Taj ul-Islam. 1992. *Pakistan as a Peasant Utopia*. Boulder, CO: Westview Press.
Hasina, Sheikh. 2015. "2015 Is Going to Be a Milestone in World History." Op-Ed. *Daily Star* (Bangladesh), January 11, 2015.
Heinig, R. L. 1892. *Working Plan of the Sunderban Government Forests: Khulna and 24-Parganas Districts, Bengal*. Calcutta: Conservator of Forests, Bengal.
Hennessy, Elizabeth. 2013. "Producing 'Prehistoric' Life: Conservation Breeding and the Remaking of Wildlife Genealogies." *Geoforum* 49: 71–80.
Herz, Barbara K. 1984. Official Development Assistance for Population Activities: A Review. World Bank (Washington, DC).
Hossain, E., S. M. Nurun Nabi, and A. Kaminksi. 2015. Climate-Smart House: Housing That Is Cyclone Resistant and Food, Energy and Water Efficient in Bangladesh. WorldFish (Penang, Malaysia).
Hossain, Faisal, Zahirul Haque Khan, and C. K. Shum. 2015. "Tidal River Management in Bangladesh." *Nature Climate Change* 5 (6): 492.
Hossain, Naomi. 2017. *The Aid Lab: Understanding Bangladesh's Unexpected Success*. Oxford: Oxford University Press.
Hossen, M. Anwar, and John R. Wagner. 2015. "The Need for Community Inclusion in Water Basin Governance in Bangladesh." *Bandung: Journal of the Global South* 2: 1–18.
Howell, Junia, and James R. Elliott. 2019. "Damages Done: The Longitudinal Impacts of Natural Hazards on Wealth Inequality in the United States." *Social Problems* 66: 448–467.
Human Rights Watch. 2015. "Marry before Your House Is Swept Away: Child Marriage in Bangladesh." Human Rights Watch.
Hummelbrunner, Richard. 2010. "Beyond Logframe: Critique, Variations and Alternatives." In *Beyond Logframe: Using Systems Concepts in Evaluation*, ed. N. Fujita, 1–34. Tokyo: Foundation for Advanced Studies on International Development.
Hunter, W. W. 1875a. *A Statistical Account of Bengal. Vol. 1: Districts of the 24 Parganas and Sundarbans*. London: Trübner & Co.
———. 1875b. *A Statistical Account of Bengal. Vol. 2: Districts of Nadiyá and Jessor*. London: Trübner & Co.
Huq, A. M. 1958. "Reflections on Economic Planning in Pakistan." *Far Eastern Survey* 27 (1): 1–6.
Huq, Saleemul, Syed Iqbal Ali, and A. Atiq Rahman. 1995. "Sea-Level Rise and Bangladesh: A Preliminary Analysis." *Journal of Coastal Research* 14: 44–53.
International Bank for Reconstruction and Development. 1970. "Proposals for an Action Program: East Pakistan Agriculture and Water Development." July 17, 1970.
———. 1972a. "Crops, Livestock and Fisheries," December 1, 1970.
———. 1972b. "Detailed Sector Review," December 1, 1970.
———. 1972c. "The Regional Development Potentials and Constraints," Decemer 1, 1972.
———. 1972d. "Water."
IOR. 1915. "Bengal Embankments (Sundarbans) Law 1915." IOR/L/PJ/6/1347, File 229. India Office Records and Private Papers. London: British Library.
IPCC. 2013. "Summary for Policymakers." In *Climate Change 2013: The Physical Science Basis. Contribution of Working Group I to the Fifth Assessment Report of the Intergovernmental Panel on Climate Change*, edited by T. F. Stocker, D. Qin, G.-K. Plattner, M. Tignor, S. K. Allen, J. J. Boschung, A. Nauels, Y. Xia, V. Bex, and P. M. Midgley, 3–29. Cambridge: Cambridge University Press.
Iqbal, Iftekhar. 2010. *The Bengal Delta: Ecology, State and Social Change, 1840–1943*. New York: Palgrave Macmillan.

———. 2019. "Governing Mass Migration to Dhaka: Revisiting Climate Factors." *Economic and Political Weekly* 54 (36): 26–31.

Ireland, Philip, and Katharine McKinnon. 2013. "Strategic Localism for an Uncertain World: A Postdevelopment Approach to Climate Change Adaptation." *Geoforum* 47: 158–166.

Isenberg, Nancy. 2016. *White Trash*. New York: Viking.

Islam, Md. Saidul. 2008. "From Sea to Shrimp Processing Factories in Bangladesh: Gender and Employment at the Bottom of a Global Commodity Chain." *Journal of South Asian Development* 3 (2): 211–236.

Islam, Md. Shahidul, and Mahfuzul Haque. 2004. "The Mangrove-Based Coastal and Nearshore Fisheries of Bangladesh: Ecology, Exploitation and Management." *Reviews in Fish Biology and Fisheries* 14 (2): 153–180.

Islam, Shahidul, and Zakir Kibria. 2006. *Unraveling KJDRP: ADB Financed Project of Mass Destruction in Southwest Coastal Region of Bangladesh*. Dhaka: Uttaran.

Ito, S. 2002. "From Rice to Prawns: Economic Transformation and Agrarian Structure in Rural Bangladesh." *Journal of Peasant Studies* 29 (2): 47–70.

Jasanoff, Sheila. 2004. "The Idiom of Co-production." In *States of Knowledge: The Co-production of Science and Social Order*, edited by Sheila Jasanoff, 1–12. London: Routledge.

———. 2010. "A New Climate for Society." *Theory, Culture & Society* 27 (2–3): 233–253.

———. 2015. "Imagined and Invented Worlds." In *Dreamscapes of Modernity: Sociotechnical Imaginaries and the Fabrication of Power*, edited by Sheila Jasanoff and Sang-Hyun Kim, 321–340. Chicago: University of Chicago Press.

Jasanoff, Sheila, and Sang-Hyun Kim, eds. 2015. *Dreamscapes of Modernity: Sociotechnical Imaginaries and the Fabrication of Power*. Chicago: University of Chicago Press.

Joffre, O, M Prein, PBV Tung, SB Saha, NV Hao, and MJ Alam. 2010. "Evolution of Shrimp Aquaculture Systems in the Coastal Zones of Bangladesh and Vietnam: a Comparison." In *Tropical Deltas and Coastal Zones: Food Production, Communities and Environment at the Land-Water Interface*, edited by Chu T Hoanh, Brian Szuster, Kam Suan-Pheng, Abdelbagi M Ismail and Andrew D Noble, 48–63. Oxfordshire: CABI.

Karim, Lamia. 2011. *Microfinance and its Discontents: Women in Debt in Bangladesh*. Minneapolis: University of Minnesota Press.

Karim, M, Manjur Murshed Zahid Ahmed, RK Talukder, MA Taslim, and HZ Rahman. 2006. *Policy Working Paper: Dynamic Agribusiness-Focused Aquaculture for Poverty Reduction and Economic Growth in Bangladesh*. WorldFish Center (Penang, Malaysia).

Kates, Robert W., William R. Travis, and Thomas J. Wilbanks. 2012. "Transformational Adaptation when Incremental Adaptations to Climate Change Are Insufficient." *Proceedings of the National Academy of Sciences of the United States of America* 109 (19): 7156–7161.

Katz, Cindi. 1995. "Under the Falling Sky: Apocalyptic Environmentalism and the Production of Nature." In *Marxism in the Postmodern Age: Confronting the New World Order*, edited by Antonio Callari, Stephen Cullenberg, and Carole Biewener, 276–282. New York: Guilford Press.

Kautsky, Karl. 1988 [1899]. *The Agrarian Question, Volume 1*. London: Zwan Publications.

Kay, S., J. Caesar, Judith Wolf, L. Bricheno, R. J. Nicholls, A. K. M. Saiful Islam, A. Haque, A. Pardaens, and J. A. Lowe. 2015. "Modelling the Increased Frequency of Extreme Sea Levels in the Ganges-Brahmaputra-Meghna Delta Due to Sea Level Rise and Other Effects of Climate Change." *Environmental Science: Processes & Impacts* 17: 1311–1322.

Kerr, Ron. 2008. "International Development and the New Public Management: Projects and Logframes as Discursive Technologies of Governance." In *The New Development Management: Critiquing the Dual Modernization*, edited by Sadhvi Dar and Bill Cooke, 91–110. London: Zed Books.
Khadim, Fahad Khan, Kanak Kanti Kar, Pronab Kumar Halder, Md. Atiqur Rahman, and A. K. M. Mostafa Morshed. 2013. "Integrated Water Resources Management (IWRM) Impacts in South West Coastal Zone of Bangladesh and Fact-Finding on Tidal River Management." *Journal of Water Resource and Protection* 5 (10): 953–961.
Khan, Akhter Hameed. 1974. *Reflections on the Comilla Rural Development Projects*. Overseas Liaison Committee.
Khan, Arastoo. 2013. "Bangladesh—The Most Climate Vulnerable Country." *End Poverty in South Asia* (blog). World Bank (Washington, DC).
Khan, Azizur Rahman. 1979. "The Comilla Model and the Integrated Rural Development Programme of Bangladesh An Experiment in 'Cooperative Capitalism.'" *World Development* 7: 397–422.
Khandker, Shahidur R., Gayatri B. Koolwal, and Syed Badruddoza. 2013. "How Does Competition Affect the Performance of MFIs?" World Bank (Washington, DC).
Khanom, Tanzinia. 2016. "Living on the Edge." *Dhaka Tribune*, September 17, 2016. Climate Change section.
Khulna University. 2014. "Report on Baseline Survey and Socioeconomic Analysis of the Project Area (Tala, Satkhira)." Khulna University and UNDP.
Kipling, Rudyard. 1922. *Rudyard Kipling's Verse*. Garden City, NY: Doubleday, Page & Company.
Klein, Naomi. 2008. *The Shock Doctrine*. New York: Picador.
———. 2014. *This Changes Everything: Capitalism vs. The Climate*. New York: Simon & Schuster.
Knorr-Cetina, Karin. 1992. "The Couch, the Cathedral, and the Laboratory: On the Relationship between Experiment and Laboratory in Science." In *Science as Practice and Culture*, edited by Andrew Pickering, 113–138. Chicago: University of Chicago Press.
Koch, Fred G. 1991. "Cyclone and Coastal Protection." Paper presented at the conference Bangladesh Disaster: Issues and Perspectives, Delft, the Netherlands, September 3.
Koslov, Liz. 2016. "The Case for Retreat." *Public Culture* 28 (2): 359–387.
———. 2019. "Avoiding Climate Change: 'Agnostic Adaptation' and the Politics of Public Silence." *Annals of the American Association of Geographers* 109 (2): 568–580.
Kroodsma, David. 2015. "The Delta." *Ride for Climate* (blog). http://rideforclimate.com/blog/?p=1300.
Lahiri-Dutt, Kuntala. 2014. "Commodified Land, Dangerous Water: Colonial Perceptions of Riverine Bengal." In "Asian Environments: Connections across Borders, Landscapes, and Times," edited by Ursula Münster, Shiho Satsuka, and Gunnel Cederlöf. Special issue, *RCC Perspectives*, 2014/3: 17–22.
Lahiri-Dutt, Kuntala, and Gopa Samanta. 2013. *Dancing with the River: People and Life on the Chars of South Asia*. New Haven, CT: Yale University Press.
Landau, Lauren B. 2012. "Communities of Knowledge or Tyrannies of Partnership: Reflections on North-South Research Networks and the Dual Imperative." *Journal of Refugee Studies* 25 (4): 555–570.
Lázár, Attila N., Derek Clarke, Helen Adams, Abdur Razzaque Akanda, Sylvia Szabo, Robert J. Nicholls, Zoe Matthews, Dilruba Begum, Abul Fazal M. Saleh, Md. Anwarul Abedin, Andres Payo, Peter Kim Streatfield, Craig Hutton, M. Shahjahan Mondal, and Abu Zofar Md. Moslehuddin. 2015. "Agricultural Livelihoods in Coastal Bangladesh

under Climate and Evironmental Change—A Model Framework." *Environmental Science: Processes & Impacts* 17: 1018–1031.
Leedshill–De Leuw Engineers. 1968. "Coastal Embankment Project Engineering and Economic Evaluation." East Pakistan Water and Power Development Authority (Dacca, East Pakistan).
Lees, O. C. 1906. *Waterways in Bengal: Their Economic Value and the Methods Employed for Their Improvement*. Calcutta: Bengal Secretariat Book Depot.
Levien, Michael. 2018. *Dispossession without Development: Land Grabs in Neoliberal India*. New York: Oxford University Press.
Levitas, Ruth. 2010. *The Concept of Utopia*. Bern: Peter Lang AG.
Lewis, David. 2010. "The Strength of Weak Ideas? Human Security, Policy History, and Climate Change in Bangladesh." In *Security and Development*, edited by John-Andrew McNeish and Jon Harald Sande Lie, 113–129. Oxford: Berghahn Books.
———. 2011. *Bangladesh: Politics, Economy and Civil Society*. Cambridge: Cambridge University Press.
———. 2014. "Commodifying Development Experience: Deconstructing Development as Gift in the Development Blockbuster." *Anthropological Forum* 24 (4): 440–453.
———. 2017. "Organising and Representing the Poor in a Clientelistic Democracy: the Decline of Radical NGOs in Bangladesh." *Journal of Development Studies* 53 (10): 1545–1567.
———. 2019. "Akhtar Hameed Khan (1914–1999)." In *Key Thinkers on Development*, edited by David Simon, 239–248. Oxon, UK: Routledge.
Lewis, David, and Abul Hossain. 2019. "Local Political Consolidation in Bangladesh: Power, Informality and Patronage." *Development and Change*, July 8, 2019. https://doi.org/10.1111/dech.12534.
Lewis, David, and Mark Schuller. 2017. "Engagement with a Productively Unstable Category: Anthropologists and Nongovernmental Organizations." *Current Anthropology* 58 (5): 634–651.
Li, Tania Murray. 2007. *The Will to Improve: Governmentality, Development, and the Practice of Politics*. Durham, NC: Duke University Press.
———. 2014. *Land's End: Capitalist Relations on an Indigenous Frontier*. Durham, NC: Duke University Press.
Loftus, Alex. 2009a. "Intervening in the Environment of the Everyday." *Geoforum* 40: 326–334.
———. 2009b. "The *Theses on Feuerbach* as a Political Ecology of the Possible." *Area* 41 (2): 157–166.
———. 2013. "Gramsci, Nature, and the Philosophy of Praxis." In *Gramsci: Space, Nature, Politics*, edited by Michael Ekers, Gillian Hart, Stefan Kipfer, and Alex Loftus, 178–196. Malden, MA: Wiley-Blackwell.
Lohmann, Larry. 2011. "Capital and Climate Change." *Development and Change* 42 (2): 649–668.
Ludden, David. 2005. "Development Regimes in South Asia: History and the Governance Conundrum." *Economic and Political Weekly* 40 (37): 4042–4051.
———. 2006. "A Useable Past for a Post-national Present: Governance and Development in South Asia." *Journal of the Asiatic Society of Bangladesh* 50 (1–2): 259–292.
———. 2011. *An Agrarian History of South Asia*. Cambridge: Cambridge University Press.
Mackay, James Aberigh. 1860. *From London to Lucknow*. London: Jamest Nisbet and Co.
Mackill, David J., Abdelbagi M. Ismail, Alvaro M. Pamplona, Darlene L. Sanchez, Jerome J. Carandang, and Endang M. Septiningsih. 2010. "Stress Tolerant Rice Varieties for Adaptation to a Changing Climate." *Crop, Environment & Bioinformatics* 7: 250–259.

Mahmud, Muhammad Shifuddin, Dik Roth, and Jeroen Warner. 2020. "Rethinking 'Development': Land Dispossession for the Rampal Power Plant in Bangladesh." *Land Use Policy* 94: 104492.

Maiti, Ramkrishna, Tapan Kumar Das, and Animesh Majee. 2010. "Cognition of the Interworking of Processes Leading to Embankment Breaching along the Raimangal, Sundarbans." *Water and Energy International* 67 (1): 25–34.

Maitra, Bisweswar. 1972. "Sundarbans Delta Project and Its Role in the Economic Development of West Bengal." In *Focus on West Bengal: Problems and Prospects*, edited by Dhires Bhattacharyya, 126–137. Calcutta: Arghya Kusum Datta Gupta.

Mandal, Asim Kumar. 2003. *The Sundarbans of India: A Development Analysis*. New Delhi: Indus Publishing.

Marino, Elizabeth. 2018. "Adaptation Privilege and Voluntary Buyouts: Perspectives on Ethnocentrism in Sea Level Rise Relocation and Retreat Policies in the US." *Global Environmental Change* 49: 10–13.

Marino, Elizabeth, and Jesse Ribot. 2012. "Special Issue Introduction: Adding Insult to Injury: Climate Change and the Inequities of Climate Intervention." *Global Environmental Change* 22: 323–328.

Marshall, P. J. 1987. *Bengal: The British Bridgehead, Eastern India 1740–1828*. Cambridge: Cambridge University Press.

Martinez-Alier, Joan. 2002. *The Environmentalism of the Poor: A Study of Ecological Conflicts and Valuation*. Cheltenham, UK: Edward Elgar.

Marx, Karl, and Frederich Engels. (1846) 1998. *The German Ideology*. Amherst, NY: Prometheus Books.

McAdam, Doug, Sidney Tarrow, and Charles Tilly. 2001. *Dynamics of Contention*. Cambridge: Cambridge University Press.

McCarthy, James. 2015. "A Socioecological Fix to Capitalist Crisis and Climate Change? The Possibilities and Limits of Renewable Energy." *Environment and Planning A* 47: 1–18.

McMichael, Philip. 1990. "Incorporating Comparison within a World-Historical Perspective: An Alternative Comparative Method." *American Sociological Review* 55 (3): 385–397.

———. 2004. *Development and Social Change: A Global Perspective*. 3rd ed. Thousand Oaks, CA: Pine Forge Press.

———. 2006. "Peasant Prospects in the Neoliberal Age." *New Political Economy* 11 (3): 407–418.

———. 2008. "Peasants Make Their Own History, But Not Just as They Please . . ." *Journal of Agrarian Change* 8 (2–3): 205–228.

———. 2009. "Contemporary Contradictions of the Global Development Project: Geopolitics, Global Ecology and the 'Development Climate.'" *Third World Quarterly* 30 (1): 247–262.

———. 2013. *Food Regimes and Agrarian Questions*. Halifax, Canada: Fernwood Publishing.

Mifflin, Alex. 2013. "Bangladesh Is Drowning Because of Climate Change." *HuffPost Impact Canada* (blog). https://www.huffingtonpost.ca/alex-mifflin/bangladesh-climate-change_b_4150220.html.

Miller, Clark A. 2004. "Climate Science and the Making of a Global Political Order." In *States of Knowledge: The Co-production of Science and Social Order*, edited by Sheila Jasanoff, 46–66. New York: Routledge.

Miller, Clark A., and Paul Edwards, eds. 2001. *Changing the Atmosphere: Expert Knowledge and Environmental Governance*. Cambridge: MIT Press.

Ministry of Agriculture, Bangladesh, and FAO. 2013. "Master Plan for Agricultural Development in the Southern Region of Bangladesh." Ministry of Agriculture, Government of the People's Republic of Bangladesh and Food and Agriculture Organization of the United Nations (Dhaka).

Ministry of Water Resources, Bangladesh. 2001. *Integrated Environmental Management: A Case Study on Shrimp-Paddy Land Use Strategies in the Southwest of Bangladesh.* Edited by Environment and GIS Support Project for Water Sector Planning. Dhaka: Ministry of Water Resources, Government of Bangladesh.

Mitchell, Timothy. 1988. *Colonising Egypt.* Berkeley: University of California Press.

——. 2002. *Rule of Experts: Egypt, Techno-Politics, and Modernity.* Berkeley: University of California Press.

Mitra, Iman Kumar. 2016. "Migrant Workers and Informality in Contemporary Kolkata." In *Cities, Rural Migrants and the Urban Poor: Issues of Violence and Social Justice, Research Briefs with Policy Implications,* 14–17. Kolkata: Mahanirban Calcutta Research Group.

MoEF. 2009. *Bangladesh Climate Change Strategy and Action Plan 2009.* Dhaka: Ministry of Environment and Forests, Government of the People's Republic of Bangladesh.

Moni, Sonia H. 2014. "Black Tiger Supply Crunch Pushes Up Shrimp Prices." *Financial Express Bangladesh,* September 24, 2014.

Mottaleb, Khondoker Abdul, and Tetsushi Sonobe. 2011. "An Inquiry into the Rapid Growth of the Garment Industry in Bangladesh." *Economic Development and Cultural Change* 60 (1): 67–89.

Mozena, Dan. 2014. "The New Bangladesh." *AmCham: Journal of the American Chamber of Commerce in Bangladesh* 7 (4): 13–16.

Muhammad, Anu. 2006. "Globlisation and Economic Transformation in a Peripheral Economy: The Bangladesh Experience." *Economic and Political Weekly* 41 (15): 1459–1464.

Mukerjee, Radhakamal. 1938. *The Changing Face of Bengal: A Study in Riverine Economy.* Calcutta: University of Calcutta.

Mukherjee, K. N. 1969a. "Harmonious Solution of the Basic Problem of Sundarban Reclamation." *Geographical Review of India* 38 (3): 311–315.

——. 1969b. "Nature and Problems of Neo-Reclamation in the Sundarbans." *Geographical Review of India* 31 (4): 1–9.

——. 1983. "History of Settlement in the Sundarban of West Bengal." *Indian Journal of Landscape Systems* 6 (1–2): 1–19.

Mundy, Godfrey Charles. 1858. *Journal of a Tour in India.* London: John Murray, Albemarle Street.

Muzzini, Elisa, and Gabriela Aparicio. 2013. "Bangladesh: The Path to Middle-Income Status from an Urban Perspective." World Bank (Washington, DC).

Nakao, Takehiko, and Yvo de Boer. 2015. "Climate Change and a Prosperous Asia." Op-Ed. *Dhaka Tribune,* January 9, 2015.

Nandy, Subinay. 1991. "Socio-Economic and Demographic Aspects of Natural Disasters in Bangladesh." Paper presented at the conference Bangladesh Disaster: Issues and Perspectives, Delft, the Netherlands, September 3.

Nasiruddin, Mohammed, and Lia Carol Sieghart. 2014. "The Hot Spot." Op-Ed. *Dhaka Tribune,* March 15, 2014.

National Research Council. 1971. "East Pakistan Land and Water Development as Related to Agriculture." National Academy of Sciences, Board on Science and Technology for International Development (Washington, DC).

Netherlands Ministry of Foreign Affairs. 1996. "Bangladesh: Evaluation of the Netherlands Development Programme with Bangladesh, 1972–1996." Netherlands Ministry of Foreign Affairs.

Newell, Peter, and Olivia Taylor. 2018. "Contested Landscapes: The Global Political Economy of Climate-Smart Agriculture." *Journal of Peasant Studies* 45 (1): 108–129.

New York Times. 2008. "Dutch Draw up Drastic Measures to Defend Coast against Rising Seas." September 3, 2008.

Nicholls, Robert J., and Steven L. Goodbred Jr. 2004. "Towards Integrated Assessment of the Ganges-Brahmaputra Delta." Paper presented at the 5th International Conference on Asian Marine Geology, Seoul, South Korea, October 16–18.

Nicholls, Robert J., Craig Hutton, Attila N. Lázár, Md. Munsur Rahman, Mashfique Salehin, and Tuhin Ghosh. 2013. "Understanding Climate Change and Livelihoods in Coastal Bangladesh." *Hydrolink* 2: 40–42.

Nicholls, Robert J., Paul Whitehead, Judith Wolf, Munsur Rahman, and Mashfiqus Salehin. 2015. "The Ganges-Brahmaputra-Meghna Delta System: Biophysical Models to Support Analysis of Ecosystem Services and Poverty Alleviation." *Environmental Science: Processes & Impacts* 17: 1016–1017.

Nightingale, Andrea J. 2017. "Power and Politics in Climate Change Adaptation Efforts: Struggles over Authority and Recognition in the Context of Political Instability." *Geoforum* 84: 11–20.

Noble, I. R., S. Huq, Y. A. Anokhin, J. Carmin, D. Goudou, F. P. Lansigan, B. Osman-Elasha, and A. Villamizar. 2014. "Adaptation Needs and Options." In *Climate Change 2014: Impacts, Adaptation, and Vulnerability. Part A: Global and Sectoral Aspects. Contribution of Working Group II to the Fifth Assessment Report of the Intergovernmental Panel on Climate Change*, edited by C. B. Field, V. R. Barros, D. J. Dokken, K. J. Mach, M. D. Mastrandrea, T. E. Bilir, M. Chatterjee, K. L. Ebi, Y. O. Estrada, R. C. Genova, B. Girma, E. S. Kissel, A. N. Levy, S. MacCracken, P. R. Mastrandrea, and L. L. White, 833–868. Cambridge: Cambridge University Press.

Nuruzzaman, M. 2006. "Dynamics and diversity of shrimp farming in Bangladesh: technical aspects." In *Shrimp Farming and Industry: sustainability, trade and livelihoods*, edited by A Atiq Rahman, AHG Quddus, Bob Pokrant, and Md Liaquat Ali, 431–460. Dhaka: University Press Limited.

O'Brien, Karen. 2012. "Global Environmental Change II: From Adaptation to Deliberate Transformation." *Progress in Human Geography* 36 (5): 667–676.

———. 2013. "Global Environmental Change III: Closing the Gap between Knowledge and Action." *Progress in Human Geography* 37 (4): 587–596.

O'Malley, L. S. S. 1908. *Bengal District Gazetteers: Khulna*. Calcutta: Bengal Secretariat Book Depot.

O'Reilly, Jessica. 2017. *The Technocratic Antarctic: An Ethnography of Scientific Expertise and Environmental Governance*. Ithaca, NY: Cornell University Press.

Osborne, Tracey Muttoo. 2011. "Carbon Forestry and Agrarian Change: Access and Land Control in a Mexican Rainforest." *Journal of Peasant Studies* 38 (4): 859–883.

Ovi, Ibrahim Hossain. 2014. "Frozen Food Export Earnings from 17% in FY14." *Dhaka Tribune*, July 11, 2014. Business section.

Oxfam. 2010. "A Tale of Two Worlds." *Oxfam Asia* (blog). April 21, 2010. http://www.oxfamblogs.org/asia/a-tale-of-two-worlds/.

Paprocki, Kasia. 2016. "'Selling Our Own Skin:' Social Dispossession through Microcredit in Rural Bangladesh." *Geoforum* 74: 29–38.

———. 2019. "All That Is Solid Melts into the Bay: Anticipatory Ruination and Climate Change Adaptation." *Antipode* 51 (1): 295–315.

———. 2020. "The Climate Change of Your Desires: Climate Migration and Imaginaries of Urban and Rural Climate Futures." *Environment and Planning D: Society and Space* 38 (2): 248–266.

Paprocki, Kasia, and Jason Cons. 2014. "Life in a Shrimp Zone: Aqua- and Other Cultures of Bangladesh's Coastal Landscape." *Journal of Peasant Studies* 41 (6): 1109–1130.

Pargiter, Frederick Eden. 1934. *A Revenue History of the Sundarbans from 1765 to 1870.* Alipore, Bengal: Bengal Government Press.

Parvez, Sohel. 2018. "Shrimp Exports Fall for Fourth Straight Year." *Daily Star* (Bangladesh), July 23, 2018. Business section.

Paul, Ruma. 2020. "Garment exporter Bangladesh faces $6 billion hit as top retailers cancel." *Reuters*, March 31, 2020. Business News section.

Peet, Richard, and Michael Watts, eds. 2004. *Liberation Ecologies: Environment, Development, Social Movements.* 2nd ed. London: Routledge.

Pelling, Mark. 2011. *Adaptation to Climate Change: From Resilience to Transformation.* London: Routledge.

Peluso, Nancy Lee. 1992. *Rich Forests, Poor People: Resource Control and Resistance in Java.* Berkeley: University of California Press.

Pethick, John, and Julian D. Orford. 2013. "Rapid Rise in Effective Sea-Level in Southwest Bangladesh: Its Causes and Contemporary Rates." *Global and Planetary Change* 111: 237–245.

Phillimore, R. H. 1945. *Historical Records of the Survey of India.* Behra Dun, India: Surveyor General of India.

Piddington, Henry. 1853. *A Letter to the Most Noble James Andrew, Marquis of Dalhousie, Governor General of India, on the Storm-Waves of the Cyclones in the Bay of Bengal, and their Effects in the Sunderbunds.* Calcutta: Baptist Mission Press.

Pokrant, Bob. 2014. "Brackish Water Shrimp Farming and the Growth of Aquatic Monocultures in Coastal Bangladesh." In *Historical Perspectives of Fisheries Exploitation in the Indo-Pacific*, edited by Joseph Christensen and Malcolm Tull, 107–132. Dordrecht: Springer Netherlands.

Primavera, J. H. 1997. "Socio-Economic Impacts of Shrimp Culture." *Aquaculture Research* 28: 815–827.

Prinsen, Gerard, and Saskia Nijhof. 2015. "Between Logframes and Theory of Change: Reviewing Debates and a Practical Experience." *Development in Practice* 25 (2): 234–246.

Prudham, Scott. 2009. "Pimping Climate Change: Richard Branson, Global Warming, and the Performance of Green Capitalism." *Environment and Planning A* 41 (7): 1594–1613.

Pulido, Laura. 2018. "Racism and the Anthropocene." In *Future Remains: A Cabinet of Curiosities for the Anthropocene*, edited by Gregg Mitman, Marco Armiero, and Robert S. Emmett, 116–128. Chicago: University of Chicago Press.

Quddus, A. H. G. 2006. "Livelihood of Shrimp Fry Collectors in Bangladesh." In *Shrimp Farming and Industry: Sustainability, Trade and Livelihoods*, edited by A. Atiq Rahman, A. H. G. Quddus, Bob Pokrant, and Md. Liaquat Ali, 355–367. Dhaka: University Press Limited.

Rademacher, Anne. 2018. *Building Green: Environmental Architects and the Struggle for Sustainability in Mumbai.* Oakland: University of California Press.

Rahman, A. Atiq, Zahid H. Chowdhury, and Ahsan U. Ahmed. 2003. "Environment and Security in Bangladesh." In *Environment, Development, and Human Security: Perspectives from South Asia*, edited by Adil Najam, 103–128. Lanham, MD: University Press of America.

Rahman, Atiur. 1995. *Beel Dakatia: The Environmental Consequences of a Development Disaster.* Dhaka: University Press Limited.

Rahman, Mohammed Ataur, and Sowmen Rahman. 2015. "Natural and Traditional Defense Mechanisms to Reduce Climate Risks in Coastal Zones of Bangladesh." *Weather and Climate Extremes* 7: 84–95.

Rainey, H. James. 1868. "What Was the Sundarban Originally, and When, and Wherefore Did It Assume Its Existing State of Utter Desolation?" *Proceedings of the Asiatic Society of Bengal, January to December 1868*: 264–273.

———. 1872. "The Lost City." *Mookerjee's Magazine* 1 (1–5): 343–350.

———. 1874. "Famines in Bengal, and the Reclamation of the Sundarban as a Means of Mitigating Them." *Calcutta Review* 59 (118): 332–349.

———. 1891. "The Sundarban: Its Physical Features and Ruins." *Proceedings of the Royal Geographical Society and Monthly Record of Geography* 13 (5): 273–287.

———. 1897. "The Earthquake of 1737." In *The Earthquake in Bengal and Assam: Reprinted from the "Englishman,"* edited by S. N. Banerji, 314–316. Calcutta: "Englishman" Press.

Rangan, Haripriya. 2000. *Of Myths and Movements: Rewriting Chipko into Himalayan History*. London: Verso.

Ranganathan, Malini, and Eve Bratman. 2021. "From Urban Resilience to Abolitionist Climate Justice." *Antipode* 53 (1): 115–137.

Rashid, Mamun, and Dilshad Rahat Ara. 2015. "Modernity in Tradition: Reflections on Building Design and Technology in the Asian Vernacular." *Frontiers of Architectural Research* 4: 46–55.

Razavi, Shahra. 2009. "Engendering the Political Economy of Agrarian Change." *Journal of Peasant Studies* 36 (1): 197–226.

Rennell, James. 1781. *A Bengal Atlas: Containing Maps of the Theatre of War and Commerce on That Side of Hindoostan*. London: n.p.

———. 1788. *Memoir of a Map of Hindoostan*. 2nd ed. London: M. Brown.

———. 1910. *The Journals of Major James Rennell, First Surveyor-General of India*, edited by T. H. D. La Touche. Calcutta: Asiatic Society.

Rice, Jennifer L., Brian J. Burke, and Nik Heynen. 2015. "Knowing Climate Change, Embodying Climate Praxis: Experiential Knowledge in Southern Appalachia." *Annals of the Association of American Geographers* 105 (2): 253–262.

Richards, J. F., and E. P. Flint. 1990. "Long-term Transformations in the Sundarbans Wetlands Forests of Bengal." *Agriculture and Human Values* 7 (2): 17–33.

Rickards, L., and S. M. Howden. 2012. "Transformational Adaptation: Agriculture and Climate Change." *Crop & Pasture Science* 63: 240–250.

Rippke, Ulrike, Julian Ramirez-Villegas, Andy Jarvis, Sonja J. Vermeulen, Louis Parker, Flora Mer, Bernd Diekkrüger, Andrew J. Challinor, and Mark Howden. 2016. "Timescales of Transformational Climate Change Adaptation in Sub-Saharan African Agriculture." *Nature Climate Change* 6: 605–610.

Rist, Gilbert. 2007. "Development as a Buzzword." *Development in Practice* 17 (4): 485–491.

Rochmyaningsih, Dyna. 2018. "Showcase Scientists from the Global South." *Nature* 553 (251).

Rogaly, Ben. 1996. "Micro-finance Evangelism, 'Destitute Women,' and the Hard Selling of a New Anti-poverty Formula." *Development in Practice* 6 (2): 100–112.

Rogelji, Joeri, Michel den Elzen, Niklas Höhne, Taryn Fransen, Hanna Fekete, Harald Winkler, Roberto Schaeffer, Fu Sha, Keywan Riahi, and Malte Meinshausen. 2016. "Paris Agreement Climate Proposals Need a Boost to Keep Warming Well Below 2°C." *Nature* 534: 631–639.

Rogers, Kimberly G., Steven L. Goodbred Jr., and Dhiman R. Mondal. 2013. "Monsoon Sedimentation on the 'Abandoned' Tide-influenced Ganges-Brahmaputra Delta Plain." *Estuarine, Coastal and Shelf Science* 131: 297–309.

Routledge, Paul. 2011. "Translocal Climate Justice Solidarities." In *Oxford Handbook of Climate Change and Society*, edited by John S. Dryzek, Richard B. Norgaard, and David Schlosberg, 384–398. Oxford: Oxford University Press.

———. 2016. "Climate Justice: Climate Change, Resource Conflicts, and Social Justice." In *Reframing Climate Change: Constructing Ecological Geopoitics*, edited by Shannon O'Lear and Simon Dalby, 67–82. London: Routledge.

Roy, Manoj, Joseph Hanlon, and David Hulme. 2016. *Bangladesh Confronts Climate Change: Keeping Our Heads above Water*. London: Anthem Press.

Roy, Pinaki. 2014. "Weather Goes Haywire." *Daily Star* (Bangladesh), August 20, 2014. Front page.

Ruttan, Vernon. 1997. "Participation and Development." In *Institutions and Economic Development*, edited by Christopher Clague, 217–232. Baltimore: Johns Hopkins University Press.

SAFE. 2012. "Living on the Edge: Is the Minimum Wage Enough? The Struggle of the Shrimp and Fish Processing Workers in Bangladesh." Social Activities for Environment (Khulna, Bangladesh).

Saifuzzaman, Md., and Mohd. Shamsul Alam. 2010. "Changes in Physical Environment of Southwestern Coastal Bangladesh: A River Basin Approach." *Khulna University Studies*, special issue: 1–7.

Samaddar, Ranabir. 1999. *The Marginal Nation: Transborder Migration from Bangladesh to West Bengal*. New Delhi: Sage.

Sarkar, Sutapa Chatterjee. 2010. *The Sundarbans: Folk Deities, Monsters and Mortals*. New Delhi: Social Science Press.

Sarwar, Golam Mahabub Md., and Colin D. Woodroffe. 2013. "Rates of Shoreline Change along the Coast of Bangladesh." *Journal of Coastal Conservation* 17: 515–526.

Sathi, Muktasree Chakma. 2016a. "Anu Muhammad Gets Death Threat Twice." *Dhaka Tribune*, October 13, 2016.

———. 2016b. "What's Going On?" *Dhaka Tribune*, October 16, 2016.

Schlich, W. 1876. "Remarks on the Sunderbuns." *Indian Forester* 1 (1): 6–11.

Segal, Michael. 2017. "The Missing Climate Change Narrative." *South Atlantic Quarterly* 116 (1): 121–124.

Seton-Karr, W. S. 1883. "Agriculture in Lower Bengal." *Journal of the Society of Arts* 31 (1582): 419–446.

———. 1899. "Comment on The Port of Calcutta by Sir Charles Cecil Stevens, KCSI." *Journal of the Society of Arts* 47 (2430): 627–654.

Shampa, Md. Ibne Mayaz Pramanik. 2012. "Tidal River Management (TRM) for Selected Coastal Area of Bangladesh to Mitigate Drainage Congestion." *International Journal of Scientific & Technology Research* 1 (5): 1–6.

Shanin, Teodor. 1966. "The Peasantry as a Political Factor." *Sociological Review* 14 (1): 5–10.

Sharda, Jai, and Tim Buckley. 2016. "Risky and Over-Subsidised: A Financial Analysis of the Rampal Power Plant." Institute for Energy Economics and Financial Analysis, June 2016.

Shaw, Rosalind. 1989. "Living with Floods in Bangladesh." *Anthropology Today* 5 (1): 11–13.

Shearer, Christine. 2012. "The Political Ecology of Climate Change Adaptation Assistance: Alaska Natives, Displacement, and Relocation." *Journal of Political Ecology* 19 (1): 174–83.

Siddiqi, Dina. 2000. "Miracle Worker or Womanmachine? Tracking (Trans)national Realities in Bangladeshi Factories." *Economic and Political Weekly* 35 (21–22): 11–17.

Siddique, Abu Bakar. 2015. "CCDB to Build Climate Tech Park Next Year." *Dhaka Tribune*, June 26, 2015. Back page.

Siddiquee, Noore Alam, and Gofran Faroqi. 2016. "Governance of NGOs in Bangladesh: Control Mechanisms and Their Limitations." In *Public Policy and Governance in*

Bangladesh: Forty Years of Experience, edited by Nizam Ahmed, 231–244. London: Routledge.

Singh, Subrata. 2013. "Common Lands Made 'Wastelands': Making of the 'Wastelands' into Common Lands." Paper presented at the 14th Global Conference of the International Association for the Study of the Commons, Kitafuji, Japan, July 3–7.

Sivaramakrishnan, K. 1999. *Modern Forests: Statemaking and Environmental Change in Colonial Eastern India*. Stanford, CA: Stanford University Press.

Skrimshire, Stefan, ed. 2010. *Future Ethics: Climate Change and Apocalyptic Imagination*. London: Continuum.

Smith, Neil. 2008. *Uneven Development: Nature, Capital, and the Production of Space*. Athens: University of Georgia Press.

Smith, Zadie. 2014. "Elegy for a Country's Seasons." *New York Review of Books*, April 3, 2014.

Smucker, Thomas A., Ben Wisner, Adolfo Mascarenhas, Pantaleo Munishi, Elizabeth E. Wangui, Gaurav Sinha, Daniel Weiner, Charles Bwenge, and Eric Lovell. 2015. "Differentiated Livelihoods, Local Institutions, and the Adaptation Imperative: Assessing Climate Change Adaptation Policy in Tanzania." *Geoforum* 59: 39–50.

Smythies, E. A. 1925. *India's Forest Wealth*. London: Humphrey Milford.

Sobhan, Rehman. 1962. "The Problem of Regional Imbalance in the Economic Development of Pakistan." *Asian Survey* 2 (5): 31–37.

———. 1971. "Who Pays for Development?" *Pakistan Forum* 1 (4): 6–7.

———. 1982. *The Crisis of External Dependence: The Political Economy of Foreign Aid to Bangladesh*. Dhaka: University Press Limited.

Soussan, John. 2000. Institutional Development: Netherlands Support to the Water Sector 1988–1998. Working document. Ministry of Foreign Affairs, Policy and Operations Evaluation Department (the Hague, Netherlands).

South Asians for Human Rights. 2015. Report of the Fact Finding Mission to Rampal, Bangladesh. (Colombo, Sri Lanka).

Steele, Paul. 2017. "As the Climate Changes, Will Bangladesh Change Too?" *Dhaka Tribune*. Climate Change section.

Stijnen, J. W., W. Kanning, S. N. Jonkman, and M. Kok. 2014. "The Technical and Financial Sustainability of the Dutch Polder Approach." *Journal of Flood Risk Management* 7 (1): 3–15.

Stonich, Susan S., and Conner Bailey. 2000. "Resisting the Blue Revolution: Contending Coalitions Surrounding Industrial Shrimp Farming." *Human Organization* 59 (1): 23–36.

Stott, Clare, and Saleemul Huq. 2014. "Knowledge Flows in Climate Change Adaptation: Exploring Friction between Scales." *Climate and Development* 6 (4): 382–387.

Sur, Malini. 2010. "Chronicles of Repression and Resilience." In *Freedom from Fear, Freedom from Want? Re-Thinking Security in Bangladesh*, edited by Hameeda Hossain, Meghna Guhathakurta, and Malini Sur. https://www.academia.edu/34302506/Freedom_from_Fear_Freedom_from_Want.

Swingle, H. S., H. R. Schmittou, D. D. Moss, and W. A. Rogers. 1969. "Fishculture Project Report for East Pakistan." Auburn University Agricultural Experiment Station (Auburn, Alabama).

Swyngedouw, Erik. 2010. "Apocalypse Forever? Post-political Populism and the Spectre of Climate Change." *Theory, Culture & Society* 27 (2–3): 213–232.

———. 2013a. "Apocalypse Now! Fear and Doomsday Pleasures." *Capitalism Nature Socialism* 24 (1): 9–18.

———. 2013b. "The Non-Political Politics of Climate Change." *ACME: An International E-Journal for Critical Geographies* 12 (1): 1–8.

Syvitski, James P. M., Albert J. Kettner, Irina Overeem, Eric W. H. Hutton, Mark T. Hannon, G. Robert Brakenridge, John Day, Charles Vörösmarty, Yoshiki Saito, Liviu Giosan, and Robert J. Nicholls. 2009. "Sinking Deltas Due to Human Activities." *Nature Geoscience* 2: 681–686.

Tacoli, Cecilia. 2009. "Crisis or Adaptation? Migration and Climate Change in a Context of High Mobility." *Environment & Urbanization* 21 (2): 513–525.

TakePart. 2017. "Hidden Connections." Thomson Reuters Foundation. http://www.takepart.com/hidden-connections/.

Talchabhadel, Rocky, Nakagawa Hajime, Kenji Kawaike, and Md. Shibly Sadik. 2018. "Polder to De-Polder: An Innovative Sediment Management in Tidal Basin in the Southwestern Bangladesh." *DPRI Annuals*, no. 61B: 623–630.

Tanigushi, Shinkichi. 1981. "The Patni System—A Modern Origin of the 'Sub-Infeudation' of Bengal in the Nineteenth Century." *Hitotsubashi Journal of Economics* 22 (1): 32–60.

Taylor, Marcus. 2015. *The Political Ecology of Climate Change Adaptation: Livelihoods, Agrarian Change and the Conflicts of Development*. London: Routledge.

———. 2018. "Climate-Smart Agriculture: What Is It Good For?" *Journal of Peasant Studies* 45 (1): 89–107.

Temple, Sir Richard. 1880. *India in 1880*. London: John Murray.

———. 1882. *Men and Events of My Time in India*. London: John Murray.

Tenzing, Janna. 2020. "Integrating Social Protection and Climate Change Adaptation: A Review." *WIREs Climate Change* 11 (2): e626.

Thomas, John W. 1972a. "Development Institutions, Projects, and Aid in the Water Development Program of East Pakistan." US Agency for International Development.

———. 1972b. "The Development of Tubewell Irrigation in Bangladesh: An Analysis of Alternatives." US Agency for International Development.

———. 1972c. "Public Policy in the Reconstruction and Development of Rural Bangladesh." US Agency for International Development.

Thomas, Kimberley Anh. 2017a. "The Ganges Water Treaty: 20 Years of Cooperation, on India's Terms." *Water Policy* 19: 724–740.

———. 2017b. "The River-Border Complex: A Border-Integrated Approach to Transboundary River Governance Illustrated by the Ganges River and Indo-Bangladeshi Border." *Water International* 42 (1): 34–53.

Thompson, E. P. (1974) 2013. *Whigs and Hunters: The Origin of the Black Act*. London: Breviary Stuff Publications.

Tilley, Helen. 2011. *Africa as a Living Laboratory*. Chicago: University of Chicago Press.

Tompkins, Emma L., Natalie Succkall, Katharine Vincent, Rezaur Rahman, Adelina Mensah, Tuhin Ghosh, and Somnath Hazra. 2017. *Observed Adaptation in Deltas*. Deltas, Vulnerability and Climate Change: Migration and Adaptation, IDRC Project Number 107642.

Transit. 1848. *A Letter to the Shareholders of the East Indian Railway, and to the Commercial Capitalists of England and India*. London: Smith, Elder and Co.

Transparency International Bangladesh. 2015. "Rampal and Matarbari Power Projects: Governance Challenges in Environmental Impact Assessment and Land Acquisition." Transparency International Bangladesh (Dhaka, Bangladesh), April 2015.

Tsing, Anna. 2000. "Inside the Economy of Appearances." *Public Culture* 12 (1): 115–144.

Uddin, Shahzad. 2005. "Privatization in Bangladesh: The Emergence of 'Family Capitalism.'" *Development and Change* 36 (1): 157–182.

UNDP. 2014. *Initiation Plan: Increasing adaptive capacities to enhance resilience of the South-West communities*. Dhaka: United Nations Development Programme, Bangladesh.

United Nations Water Control Mission to Pakistan. 1959. *Water and Power Development in East Pakistan*. New York: United Nations.

University of Dhaka. 2014. "Physical and Environmental Assessment of Proposed Inclusive and Adaptive Tidal River Management (TRM++) in Selected Kobadak Catchment Areas." Disaster Science and Management Discipline, Earth and Environmental Sciences Faculty, University of Dhaka and UNDP (Dhaka).

Unruh, Gregory C. 2000. "Understanding Carbon Lock-in." *Energy Policy* 28 (12): 817–830.

USAID. 2009. "Adapting to Coastal Climate Change: A Guidebook for Development Planners." (Washington, DC). http://www.usaid.gov/our_work/environment/climate/.

U.S. General Accounting Office. 1971. "More Timely and Realistic Evaluation Needed for Capital Development Projects in Pakistan." Agency for International Development, United States General Accounting Office (Washington, DC).

van Ellen, W. F. T. 1991. "Floods and Floodprotection." Paper presented at the conference Bangladesh Disaster: Issues and Perspectives, Delft, the Netherlands, September 3.

Van Schendel, Willem. 1982. *Peasant Mobility*. New Delhi: Manohar.

——. 2009. *A History of Bangladesh*. Cambridge: Cambridge University Press.

van Staveren, Martijn F., Jeroen F. Warner, and M. Shah Alam Khan. 2017. "Bringing in the Tides. From Closing Down to Opening Up Delta Polders via Tidal River Management in the Southwest Delta of Bangladesh." *Water Policy* 19 (2): 147–164.

Vaughn, Sarah E. 2017a. "Disappearing Mangroves: The Epistemic Politics of Climate Adaptation in Guyana." *Cultural Anthropology* 32 (2): 242–268.

——. 2017b. "Imagining the Ordinary in Participatory Climate Adaptation." *Weather, Climate and Society* 9 (3): 533–543.

Voysey, Phil. 2015. "Bangladesh: An Example of Adapting to a Harsh Environment." *Dhaka Tribune*, August 22, 2015. Climate Change section.

Warner, Jeroen F. 2008. "The Politics of Flood Insecurity: Framing Contested River Management Projects." Wageningen Universiteit.

Watts, Michael. 1983. "On the Poverty of Theory: Natural Hazards Research in Context." In *Interpretations of Calamity: From the Viewpoint of Human Ecology*, edited by K Hewitt, 231–262. Boston: Allen & Unwin.

——. 2001. "Black Acts." *New Left Review* 9: 125–139.

——. 2002. "Chronicle of a Death Foretold: Some Thoughts on Peasants and the Agrarian Question." *Österreichische Zeitschrift für Geschichtswissenschaften* 13 (4): 22–50.

——. 2003. "Development and Governmentality." *Singapore Journal of Tropical Geography* 24 (1): 6–34.

——. 2009. "The Southern Question: Agrarian Questions of Labour and Capital." In *Peasants and Globalization: Political Economy, Rural Transformation and the Agrarian Question*, edited by A Haroon Akram-Lodhi and Cristobal Kay, 262–287. London: Routledge.

——. 2011. "Ecologies of Rule: African Environments and the Climate of Neoliberalism." In *The Deepening Crisis: Governance Challenges after Neoliberalism*, 67–92. New York: New York University Press.

——. 2015. "Now and Then: The Origins of Political Ecology and the Rebirth of Adaptation as a Form of Thought." In *The Routledge Handbook of Political Ecology*, edited by Tom Perreault, Gavin Bridge and James McCarthy, 19–50. Oxon, UK: Routledge.

Watts, Michael, and Hans G. Bohle. 1993. "The Space of Vulnerability: The Causal Structure of Hunger and Famine." *Progress in Human Geography* 17 (1): 43–67.

Webber, Sophie. 2013. "Performative Vulnerability: Climate Change Adaptation Policies and Financing in Kiribati." *Environment and Planning A* 45 (11): 2717–2733.

West, Paige. 2020. "Translations, Palimpsests, and Politics. Environmental Anthropology Now." *Ethnos* 85 (1): 118–123.

Westland, J. 1874. *A Report on the District of Jessore: Its Antiquities, Its History, and Its Commerce*. 2nd ed. Calcutta: Bengal Secretariat Press.

White, Ben, Saturnino Borras, Ruth Hall, Ian Scoones, and Wendy Wolford. 2012. "The New Enclosures: Critical Perspectives on Corporate Land Deals." *Journal of Peasant Studies* 39 (3–4): 619–647.

White, Ben, and Anirban Dasgupta. 2010. "Agrofuels Capitalism: A View from Political Economy." *Journal of Peasant Studies* 37 (4): 16.

White, Sarah. 1999. "NGOs, Civil Society, and the State in Bangladesh: The Politics of Representing the Poor." *Development and Change* 30: 307–326.

Whyte, Kyle. 2017a. "Indigenous Climate Change Studies: Indigenizing Futures, Decolonizing the Anthropocene." *English Language Notes* 55 (1–2): 153–162.

———. 2017b. "Our Ancestors' Dystopia Now: Indigenous Conservation and the Anthropocene." In *Routledge Companion to the Environmental Humanities*, edited by U. Heise, Joseph Christensen and M. Niemann, 206–215. London: Routledge.

———. 2018. "Indigenous Science (Fiction) for the Anthropocene: Ancestral Dystopias and Fantasies of Climate Change Crises." *Environment and Planning E: Nature and Space* 1 (1–2): 224–242.

———. 2020. "Too Late for Indigenous Climate Justice: Ecological and Relational Tipping Points." *WIREs Climate Change* 11 (1): e603.

Wilson, Carol A., and Steven L. Goodbred Jr. 2015. "Construction and Maintenance of the Ganges-Brahmaputra-Meghna Delta: Linking Process, Morphology, and Stratigraphy." *Annual Review of Marine Science* 7: 67–88.

Wolford, Wendy. 2004. "This Land Is Ours Now: Spatial Imaginaries and the Struggle for Land in Brazil." *Annals of the Association of American Geographers* 94 (2): 409–424.

———. 2007. "Land Reform in the Time of Neoliberalism: A Many-Splendored Thing." *Antipode* 39 (3): 550–570.

———. 2010. *This Land Is Ours Now: Social Mobilization and the Meanings of Land in Brazil*. Durham: Duke University Press.

———. 2016. "Land's End: Review." *Journal of Peasant Studies* 43 (4): 950–954.

Wolford, Wendy, Saturnino M. Borras Jr., Ruth Hall, Ian Scoones, and Ben White. 2013. "Governing Global Land Deals: The Role of the State in the Rush for Land." *Development and Change* 44 (2): 189–210.

Wolford, Wendy, and Sara Keene. 2015. "Social Movements." In *The Routledge Handbook of Political Ecology*, edited by Tom Perreault, Gavin Bridge and James McCarthy, 573–584. London: Routledge.

Wong, Poh Poh, Iñigo J. Losada, Jean-Pierre Gattuso, Jochen Hinkel, Abdellatif Khattabi, Kathleen McInnes, Yoshiki Saito, and Asbury Sallenger. 2014. "Coastal Systems and Low-lying Areas." In *Climate Change 2014: Impacts, Adaptation, and Vulnerability. Part A: Global and Sectoral Aspects. Contribution of Working Group II to the Fifth Assessment Report of the Intergovernmental Panel on Climate Change*, edited by C. B. Field, V. R. Barros, D. J. Dokken, K. J. Mach, M. D. Mastrandrea, T. E. Bilir, M. Chatterjee, K. L. Ebi, Y. O. Estrada, R. C. Genova, B. Girma, E. S. Kissel, A. N. Levy, S. MacCracken, P. R. Mastrandrea and L. L. White, 361–409. Cambridge: Cambridge University Press.

Wood, Geoffrey D. 1981. "Rural Class Formation in Bangladesh 1940–1980." *Bulletin of Concerned Asian Scholars* 13 (4): 2–17.

World Bank. 1984. "Report and Recommendation of the President of the International Development Association to the Executive Directors on a Proposed Credit of SDR 23.4 Million to the People's Republic of Bangladesh for a Second Agricultural Research Project." Report No. P-3749-BD. World Bank, March 12, 1984.

———. 2010. "Implications of Climate Change for Fresh Groundwater Resources in Coastal Aquifers in Bangladesh." World Bank (Washington, DC).
———. 2012. "World Development Report: Jobs." World Bank (Washington, DC).
———. October 2014a. "Bangladesh Development Update." World Bank (Dhaka).
———. 2014b. "Building Resilience for Sustainable Development of the Sundarbans." World Bank (Washington, DC).
———.2016. "Bangladesh Development Update." World Bank (Dhaka), April 2016.
———. 2017. *World Development Indicators 2017*. Edited by the World Bank. Washington, DC: World Bank.
WTO (World Trade Organization). 2019. *World Trade Statistical Review 2019*. World Trade Organization (Geneva).
Yunus, Muhammad. 1999. *Banker to the Poor: Micro-lending and the Battle against World Poverty*. New York: PublicAffairs.
Zeiderman, Austin. 2016. *Endangered City: The Politics of Security and Risk in Bogotá*. Durham, NC: Duke University Press.

Index

Page numbers in *italics* indicate illustrations. Page numbers appended with an italic *t* indicate tables.

Abrams, Philip, 117
accretion and erosion, 46, 84, 212n2, 212n4
Act of Permanent Settlement (1793), 26
Action-Aid Bangladesh, 185
adaptation and development regimes, 20, 52–77; Bangladesh as key site of, 56–61, 211n4; capitalist/neoliberal reform models, 57–58, 76; climate justice and, 190–97; defining and describing adaptation regimes, 6–11, 54–56; development community in Dhaka, 199–200; history of climate dystopia in Khulna and, 24, 28–32, 191–92; imagination, experimentation, and dispossession as adaptation regime processes, 7, 195 (*See also* dispossession; experimentation; imagination); incremental versus transformational adaptation, 184–89; interplay between, 9–10, 20, 52–54; political economy and, 193–94; state sovereignty affected by, 57–58. *See also* circulations of knowledge and uncertainty in development practice
adaptation to climate change. *See* climate change adaptation in Bangladesh
Adnan, Shapan, 40, 174
agriculture: animal husbandry, 132, 160, 166, 171; attitudes of farmers versus researchers toward farming life, 65–66, 74–76, 118–22, 156–57; climate change and agrarian studies, 11; depeasantization, 139–40; drivers of agrarian change, 207n5; in dystopian imaginary, 65–66; fruit trees, 1, 72, 120, 132, 162–63; subsistence production, 131, 133, 140, 174; vegetables and household gardening, 1, 72, 120, 124, 131, *160*, 161, 184; viability of, 3, 114, 123, 138, 164, 178, 180, 216n4. *See also* rice agriculture
algae skimming, 144–47, *146–48*
Ali, Tariq Omar, 60, 211n8
Allison, M. A., 84
Alpana (inhabitant of Kolanihat), 1–4
aman rice crop, 40, 124, 138, 152, 163, 174, 205
animal husbandry, 132, 160, 166, 171

Anthropocene, 207n6
anticipatory ruination, 54
aquifer depletion and salinization, 83, 131–32, 164
Arjav (inhabitant of Kolanihat), 118–21, *120*, 135, 140, 141, 149
arsenic, water contamination with, 214n10
articulation of power structures, 10
Asia Pacific Adaptation Network, 67
Asian Development Bank (ADB), 66, 213n4
Auburn University, 49
avulsion, 24–25, 209n6

Bablu (resident of Tala), 107–8
Bangladesh. *See* climate change adaptation in Bangladesh
Bangladesh Academy for Rural Development (BARD), 59, 211n7
Bangladesh Agricultural Research Institute (BARI), 178
Bangladesh Climate Change Strategy and Action Plan (2009), 73
Bangladesh Environmental Lawyers Association (BELA), 167
Bangladesh Inland Water Transportation Authority (BIWTA), 93
Bangladesh Institute of Development Studies, 88
Bangladesh Rice Research Institute (BRRI), 216n10
Bangladesh University of Engineering and Technology (BUET), IWFM at, 87–88, 212n8
Bangladesh Water Development Board (BWDB), 125, 170, 178, 210n24
Barrow, Ian J., 31
"basket case," Bangladesh viewed as, 57, 211n4
Beel Dakatia, 104, 105, 109–10
Bell, Beatson, 47
Béné, Christophe, 143–44, 213n5
Bengal Embankment Manual (1875), 42–43
Bengal Landholders' Association, 46
Benoy (inhabitant of Tilokpur), 158–61
Biharis, 153–54

245

birth control, 60, 211n8
Black, Megan, 212n5
Bloomberg, Michael, 13
Blue Gold program, 181–83
bonna flooding, 40
Boro Beel, 161
borsha flooding, 40
BRAC, 137, 185, 216n14
Brahmaputra River, 24, 30, 208n2
Brammer, Hugh, 84
British Indian Association, 46
British Raj. *See* history of climate dystopia in Khulna
Burma. *See* Myanmar

Canal Movement or Khal Andolon, 167, 176–80
canals: hydraulics, 37, 39, 42, 46, 50, 105, 106, 176–78; *khals,* 45–46; used to extend growing seasons, 124, 178; waterlogging and, 101, 180
capitalism: in adaptation and development regimes, 57–58, 76; disaster capitalism, 76; green capitalism, 194; in history of climate dystopia in Khulna, 28–32
"carbon lock-in," 217n5
Center for Population, Urbanization and Climate Change, 88
CGIAR research consortium, 49–50, 116, 178
Chakravarti, B., 46–47
char, 28, *28–29*
Chatterjee, Partha, 215n21
China: garment industry in, 71; shrimp aquaculture in, 136
Chiu, Soyee, 93–94
Choldin, Harvey M., 60
circulations of knowledge and uncertainty in development practice, 20–21, 98–117; climate migration, promotion of, 98, 107, 110; development community in Dhaka, 199–200; discourse of uncertainty, 98–100, 102, 103, 110; inevitability, sense of, 99–100; rice agriculture versus shrimp aquaculture, 98, 101, 110–12; scientific understanding of climate change and, 91–94; waterlogging and, 98, 100–110; zoning, 68, 98, 112–17
Clarke, D., 216n5
climate change adaptation in Bangladesh, 1–22, 190–97; adaptation and development regimes, 6–11, 20, 52–77 (*See also* adaptation and development regimes); colonial modes of representation, persistence of, 24; conflicting narratives of social and ecological change, 18–20; conjunctural analysis of experience of climate change, 17; dystopic imaginaries and, 2, 5, 10, 11–13, 191 (*See also* dystopic imaginaries); glossary of terms, 205; historical background, 20, 23–51 (*See also* history of climate dystopia in Khulna); incremental versus transformational adaptation, 184–89; inevitability, sense of, 4–7 (*See also* inevitability, sense of); map of Bangladesh, *xiv*; methodology of study, 199–203; political economy of, 2–6, 14–18, 192–94; "poster child" for climate-related degradation, Bangladesh as, 56–61, 91, 211n4; primacy/absence of, in competing political imaginaries, 3–4, 15–17, 55–56, 207n4; reality and currency of threat, 192, 217n5; redistribution of power and resources, as opportunity for, 21–22, 190; retreat/resettlement, politics of, 13; shrimp aquaculture and, 1–4 (*See also* resistance to shrimp aquaculture; shrimp aquaculture); uncertainty, politics and practices of, 20–21; uncertainty, politics and practices of (*See also* circulations of knowledge and uncertainty in development practice; knowledge production and politics of uncertainty); urbanization as, 12–13, 53, 75, 187. *See also specific regions and locations*
Climate Change Migration Dividend theory, 73, 74
climate crisis memoirs, as literary genre, 34, 63–67, 211n10
climate justice, 2–4, 189, 190–97
"climate mafia" in Bangladesh, 15–16, 191, 199–200
climate migration. *See* migration
Climate Smart Agriculture (CSA), 175, 216n9
Climate Smart House, 68, 186, 211n14
Climate-Resident Ecosystems and Livelihoods (CREL), 62
Coastal Embankment Project (CEP), 35, 38–43, 49
Coastal Zone Policy, 113
Cold War, 31, 57
Colombia, climate change in, 207n3, 209n8
colonial period. *See* history of climate dystopia in Khulna
colonial science, 32, 47
Comilla Model, 59–60
common sense, 3, 10–11, 54, 62, 69, 79–80, 171. *See also* Gramsci, Antonio
commons, 44, 50, 131–33, 170–71, 214n12. *See also khas* lands
Community-Based Adaptation (CBA), 212n9

Community-Based Oral Testimony (CBOT), 122, 201
Conference of Parties. *See* United Nations Framework Convention on Climate Change
conjunctural analysis of experience of climate change, 17
consultants, 37–40, 87, 108–9, 116, 174, 179–80, 181, 213n8
contraceptive technologies, 60, 211n8
Corbera, Esteve, 88
cyclones, 33, 85; Aila, 93, 94, 183, 186; Amphan, 183; Sidr, 93, 94, 183

Dacope subdistrict, 78
Daily Star, 94
Darby, Stephen E., 84
Datta, Anjan, 170, 212n16
Debjit Prachanda (*gher* businessman), 127, 152
deforestation. *See* forest resources and deforestation
Delta Development Project (DDP), 50, 170, 173–74
DEltas, Vulnerability and Climate Change: Migration and Adaptation (DECCMA) project, 68
Denzau, Gertrud and Helmut, 203
Department for International Development (DFID), United Kingdom, 55, 66, 73, 74, 87, 187, 211n2
Department of Agricultural Extension (DoAE),, Bangladesh, 111, 116–17, 164
depeasantization, 122, 139–44, 215n17
development. *See* adaptation and development regimes; circulations of knowledge and uncertainty in development practice; consultants; foreign aid/development intervention
Development and Climate Days workshop, UN Climate Change Conference (Paris, 2015), 52–53
Dexter (TV show), 63
Dhaka, preference for migrating to Kolkata versus, 150, 153–54
Dickens, Charles, 23
disaster capitalism, 76
dispossession: adaptation regime, as process in, 7–10, 70–77, 195; development strategies as driver of, 54; in history of climate dystopia in Khulna, 42–51; land tenure and, 133–39, *138*; microcredit and, 61; migration and, 72–77, 96, 202; in New Town, Kolkata, 151; political economy of climate change and, 194; Rampal power plant and, 16; rights-based approach to staying in one's home, disavowals of, 10, 75; shrimp aquaculture and, 1–4, 49–50, 71–72, 89–90, 111–12, 122, 126–30; TRM++ program and, 106–8. *See also* depeasantization
distributaries, 24, 208n5, 212n2
drainage. *See* irrigation and drainage
drinking water supply, 131–32, 135
dryland versus wetland approach, 40–41
Dumuria, 104, 161, 164, 186
Durga Puja, 153
Dutch polder system and development programs. *See* Netherlands
dystopic imaginaries, 11–13, 191; inevitability, sense of, 5, 6, 10, 19–20; inseparability of Bangladesh from, 61–67; knowledge production and politics of uncertainty, intersecting with, 93–94; Polder 22 countering, 171; shrimp aquaculture and, 2; uncertainty, politics and practices of, 21; "winners and losers" discourse, 188. *See also* history of climate dystopia in Khulna

East India Company, 25, 26, 29, 35, 36, 208n2
East Pakistan period, 37–42. *See also* Coastal Embankment Project; history of climate dystopia in Khulna; independence of Bangladesh from Pakistan
East Pakistan Water and Power Development Authority (EPWAPDA), 37–38, 49, 210n23, 214n9
Eaton, Richard, 208n4, 210n19
Economic and Social Research Council (ESRC), 87
Edelman, Marc, 143, 144
Elliott, Rebecca, 8
embankment and land reclamation: discontinuation of large-scale grants for, 210n25; dispossession, leading to, 42–43, 45–47; grassroots resistance to, 41, 210n23; history of climate dystopia in Khulna and, 35–42; maintenance issues, 41–42, 183; "mud work" in shrimp aquaculture, 149; *oshto masher badh* system, 124, 170; polder system, 39–42, 49, 87, 104, 205, 214n4, 214n9; "premature reclamation," 37, 45, 210n20; silting due to, 37, 46. *See also* indigenous water management systems
employment. *See* labor and employment
erosion and accretion, 46, 84, 212n2, 212n4
ESPA Deltas, 86–87, 89, 90, 92–93, 141
EuroConsult Mott MacDonald, 181

existential threats, 33, 38, 63–64, 78–79
experimentation: adaptation regime, as process in, 7, 67–70, 195; development laboratory, Bangladesh viewed as, 59–61, 67; history of climate dystopia in Khulna and, 35–42; inventories and checklists of, 67–68

Faaland, Just, 58
fakirs (Sufi holy men), 48
family planning, 60, 211n8
Farakka Barrage, 18, 83, 207–8n12, 212n3
farming. *See* agriculture
fertility of soil, 34–35, 214n3
fish: Polder 22, cultivation in, 174; river species, access to, 132–33; tilapia cultivation, 133
Flood Action Plan (FAP), 41, 43
flood insurance, 8, 213n7
flooding: *borsha* and *bonna* (good and bad), 40; scientific understanding of, 85; waterlogging and, 100, 101
Food and Agriculture Organization (FAO), 100–101, 112–14, 116–17, 179
Forbes magazine, 57
Ford Foundation, 59, 60
foreign aid/development intervention: coherence as category, 56; experimentation by, 67–70; involvement in Bangladesh, 57–58; lack of knowledge of Bangladesh-specific issues, 49–50, 211n28; migration as climate change adaptation strategy, encouraging, 73–75; scientific research and, 86–91, 212n5; shrimp aquaculture encouraged by, 49–50. *See also* nongovernmental organizations; *specific agencies*
forest resources and deforestation: embankments and land reclamation affecting, 36; fertility and reproductive potential of soil, historical beliefs about, 34–35; government-controlled forest management, dispossession due to shift toward, 43–45; railway network, historical development of, 30–31, 43; scientific understanding of, 85; shrimp aquaculture and river bank erosion/sedimentation, 212n2; in Sundarbans, 86, 90–91; tidal channels as means of exporting, 45–46
Forsyth, Tim, 212n9
Foucault, Michel, 212n1
Friedman, Thomas, 64

Ganges Coastal Zone, 49, 178
Ganges River: drainage through Bengal Delta, 24; Farakka Barrage and, 18, 83, 207–8n12, 212n3; Gangrall River as tidal channel of, 161; Rennell's survey of, 208n2; shipping routes, development of, 30
Gangrall River, 161
garment industry in Bangladesh, 14, 15, 53, 70–71, 75, 76, 153, 157, 215n20
gendered division of labor, 3, 141, 159, 215n2. *See also* social reproduction; women
Ghana, 68
gher owners/gher businessmen, 105–6, 126, 142
ghute, 132
Gilchrist, J. B., 23
Glennon, Robert, 64–65
Global Environment Facility (GEF), 179–80
Goldman, Michael, 212n5
Goodbred, Steven L., Jr., 84
Gopalganj, 150, 166
Gorongo (inhabitant of Kolanihat), 127, 128, 135–36, 137
Grameen Bank, 61
Gramsci, Antonio, 10, 54
green capitalism, 194
Green Revolution, 175, 216n9
groundwater depletion and salinization, 83, 131–32, 164
Guha, Ramachandra, 44

Hall, Michael C., 63–64
Hall, Stuart, 10, 193
Hamilton, Walter, 28
Harrison, Henry Leland, 42–43
Hart, Gillian, 56
Hashmi, Taj ul-Islam, 215n21
Hasina, Sheikh, 14
Hazrat Khan Jahan Ali, 208n3
health/healthcare: algae skimming, problems associated with, 146–47, *148*; conditions, capacity, and payment issues, 146–47, 214n18; contraceptive technologies, 60, 211n8; suicide attempts using chemical fertilizers, 146–47
Hindus, 124, 127, 153, 214n14, 215n21
history of climate dystopia in Khulna, 20, 23–51; accumulation and development, 28–32; Act of Permanent Settlement (1793), 26; dispossession in, 42–51; experimentation, role of, 35–42; imagination, role of, 33–35; persistence of colonial modes of representation from, 24; physical geography of region contributing to, 24–26; precolonial history of region, 208nn3–4, 210n19; temporariness and impermanence of landscape, 32–33, 43; wastelands, construction and reclamation of land as, 26–28, *28, 29*, 44, 47–48, 209nn11–12

Hossain, Naomi, 58
Huq, Saleemul, 93, 200
Hurricane Katrina, 76
Hurricane Sandy, 13
hybrid seeds, 175, 216n10, 216n12

imagination: adaptation regime, as process in, 7, 61–67, 195; history of climate dystopia in Khulna and, 33–35; inseparability of Bangladesh from climate change risk in, 61–67; political force of, 12. *See also* dystopic imaginaries
impermanence. *See* temporariness and impermanence of landscape
Imtiaz Saheb (businessman in Tilokpur), 162
incremental versus transformational adaptation, 184–89
independence of Bangladesh from Pakistan, 57, 59, 170, 214n14
India, 68
Indian Forest Act (1878), 44
indigenous seeds and seed varieties, 175, *176, 177,* 216n10
indigenous water management systems, 88, 170–71, 215n21. *See also oshto masher badh;* TRM and TRM++
inevitability, sense of, 4–7; changing/restructuring, 21–22, 196–97; conjunctural analysis of experience of climate change and, 17; in development discourse, 99–100; dystopic imaginaries and, 5, 6, 10, 19–20; Polder 22 countering, 171; scientific assumptions and, 78–81, 90–91; "winners and losers" discourse, 188
Institute of Water and Flood Management (IWFM) at BUET, 87–88, 212n8
Intergovernmental Panel on Climate Change (IPCC), 87, 88, 184
International Centre for Climate Change and Development (ICCCAD), 200
International Centre for Diarrhoeal Disease Research of Bangladesh, 88
International Engineering Company (IECO), 38
International Rice Research Institute (IRRI), 211n28
International Water Management Institute (IWMI), 178
iron, water contamination with, 214n10
irrigation and drainage: adaptation/development regimes and, 59, 210n23; circulations of knowledge and uncertainty in development practice and, 101, 104; history of climate dystopia in Khulna and, 37–39, 41–42, 46;

KJDRP (Khulna-Jessore Drainage Rehabilitation Project), 213n4; resistance to shrimp aquaculture in Tilopur/Polder 22 and, 167, 176, 178, 180, 184, 216n5; shrimp aquaculture in Kolanihat and, 123–24, 129–30, 132, 134–36, 138–39; soil salinity, flushing, 83, 164, 216n5
Islam. *See* Muslims

Johnson, Ura Alexis, 211n4
judicial harassment, 166–67

Kabir, Farah, 185
Karunamoyee Sardar, and Karunamoyee Day (in Polder 22), 171, *172*
Kautsky, Karl, 11
Khal Andolon (Canal Movement), 167, 176–80
Khan, Akhter Hameed, 59
Khan, Ayub, 59, 211n7
Khan, Yahya, 211n5
khas lands, 170, 182
Khokhon (inhabitant of Tilokpur), 164–65
Khulna City, 29, 75, 158, 161, 166
Khulna district: in dystopian imaginary, 62–67; map, *xv;* out-migration from, conflicting accounts of, 19–21; peasant social movements in, 3–4, 14; physical geography of, 24–26; Rampal Power Plant protests, 15–16; shrimp aquaculture in, 1–4; Sundarbans, relationship to, 208n1; zoning for, 115. *See also* history of climate dystopia in Khulna; Kolanihat; Sundarbans; Tilokpur; *specific polders*
Khulna Gazetteer, 29, 36
Khulna-Jessore Drainage Rehabilitation Project (KJDRP), 213n4
Kipling, Rudyard, 23
Kissinger, Henry, 211nn4–5
Klein, Naomi, 5, 76
knowledge production and politics of uncertainty, 20–21, 78–97, 192; assumptions about inevitability of environmental degradation, 78–81, 90–91; circulation of knowledge about climate change and adaptation, 91–94; climate migration and, 90, 94–97; discourse of uncertainty, 78–81; dystopic imaginaries and, 93–94; human intervention, positive and negative effects of, 80; obfuscation versus uncertainty, 80; practices of uncertainty, defined, 99–100; research community, 86–91, 200–201; "stylized" narratives of, 82; understanding and coordinating data about ecological change, 81–86

Kolanihat, 122, 123–24; depeasantization in, 139–40; job loss in, 140–44, 142*t*, 143*t*; labor and employment in shrimp aquaculture in, 141–49, 142*t*, 143*t*; land tenure in, 133–39, *138*; landscape changes in, 119, 130–33; migration from, 149–57; occupation and dispossession, 126–30; resistance to shrimp aquaculture in, 1–3; rice agriculture in, 123–24, 135, 137–39, *138*; transition from rice agriculture to shrimp aquaculture in, *125,* 125–30; water management issues, 129–30. *See also specific inhabitants by name*
Kolkata: Dhaka, preference for migrating to Kolkata versus, 150, 153–54; New Town in, migration from Polder 23 to, 150–57, *151, 154,* 202
Koslov, Liz, 8–9, 13
Kothwali Thana, 59

labor and employment: gendered division of, 3, 141, 159, 215n2; household labor, 159; job loss due to transition to shrimp aquaculture, 140–44, 142*t*, 143*t*; minimum wage in Bangladesh, 215n20; *raiyat* laborers, 37, 45; in rice agriculture, 140–42, 142*t*, 164; rickshaw-van pulling, 164–65, 216n7; in shrimp agriculture, 141–49, 142*t*, 143*t*, 165; for women, 141, 144–49
Laha, Hrishikesh, 46–47
Lamont-Doherty Earth Observatory, Columbia University, 93
land distribution inequalities in Khulna, 48–49
land grabbing, 1, 55, 111–12, 116, 125–26, 127, 151, 152, 162, 171, 209n11
Land Ministry, Bangladesh, 113–14, 116
land reclamation. *See* embankment and land reclamation
Landau, Lauren B., 212n6
landless collectives. *See* Nijera Kori; Polder 22; resistance to shrimp aquaculture; Tilokpur
Leedshill–De Leuw Engineers, 38, 40–41, 49, 210n22
Lewis, David, 15, 58, 211n4, 211n7
Li, Tanya, 103
Local Government Engineering Division (LGED), 178–79
"logical framework" (logframe) approach, 213n6
Lohani, Bindu, 66

Malaysia, climate change adaptation in, 50, 211n28
Marino, Elizabeth, 13

Marshall, P. J., 209n14
Martinez-Alier, Joan, 168
Master Plan for Agricultural Development of the Southern Region of Bangladesh, 114–15
medical care. *See* health/healthcare
Meghna River, 24, 208n2
microcredit, 60–61, 108, 137, 215n16
migration: acceptance of migrants in other countries, disagreements about, 74; by Biharis, 153–54; as climate change adaptation strategy, 72–77; desirability of, for rural inhabitants, 66, 73–74, 155; as development discourse, 98; dispossession and, 72–77, 96, 202; in dystopian imaginary, 64; historical encouragement of, 44–45; Khulna district, conflicting accounts of out-migration from, 19–21; methodology of research on, 202; microcredit loans, inability to repay, 137, 215n16; politics of retreat/resettlement, 13; for religious reasons, 124, 150, 214n14; rights-based approach to staying in one's home, disavowals of, 10, 75; of Rohingya, 210n26; rural population of Bangladesh, drop in, 211n15; scientific understanding of, 90, 94–97; seasonal, 150; shrimp aquaculture leading to, 71–72, 119, 120–21, 122, 149–57; from Tilokpur, during shrimp aquaculture period, 158, 159, 160–61; transformational adaptation and, 187
minimum wage in Bangladesh, 215n20
Mitchell, Timothy, 207n1
Mitra, Iman Kumar, 154
Modi, Narendra, 202
Mongla, Special Economic Zone in, 75
Mookerjee's Magazine, 33
Mozena, Dan, 64
Muhammad, Anu, 16
Mukerjee, Radhakamal, 210n19
Munshiganj, 62, 186
Muslims, 48, 150, 208n4, 209n15, 215n21
Myanmar (Burma): indigenous populations shifting between Bangladesh and, 210n26; Rohingya migrating to, 210n26; "wastelands" in, 45

National Aquaculture Development Strategy and Action Plan of Bangladesh, 112
National Brackish-water Research Station, Paikgachha, 123
Natural Environment Research Council (NERC), 87
Nature Climate Change, 85, 186

Netherlands: development programs in Bangladesh, 50, 87, 114, 123, 170, 173–74, 181; polder system in, 24, 35, 39–41
New Town, Kolkata, migration from Polder 23 to, 150–57, *151, 154*
New York City, adaptation regime in, 8–9, 13, 207n4
New York Times, 64, 78, 84
Nicholls, Robert J., 84, 209n7
Nijera Kori, 3–4, 15; Blue Gold program and, 182, 183; DDP and, 50, 170; in Kolanihat, 141; on labor and employment in rice agriculture versus shrimp aquaculture, 141; methodology of research and, 199, 201–2; in Polder 22, 169–71, 174, 182; seed bank, *177;* in Tilokpur, 158, 165–66
9/11, 76
Nishat, Ainun, 49–50
Nixon, Richard, 211n5
nongovernmental organizations (NGOs): adaptation/development regimes and, 58; "climate mafia" working for, 15, 16, 17, 55–56; depoliticizing impulse of, 17–18, 55; development community in Dhaka, 199–200; experimentation by, 67–70; hybrid seeds and, 175; methodology of research and, 199–200, 203; migration as climate change adaptation strategy, encouraging, 73–75; "radical NGO" sector, 15, 207n11; RTI (Right to Information Act) applicable to, 168; scientific research and, 86–91, 212n5; shrimp aquaculture and, 101, 103, 122–23, *123,* 143–44, 156, 213n5; study of climate change in Bangladesh by, 5, 14; transformational versus incremental adaptation, 186, 188; on zoning, 113, 114. *See also* foreign aid/development intervention; *specific organizations*

Obama, Barack, 64
organophosphates, 146–47, *148*
oshto masher badh, 38, 45–46, 124, 170
out-migration. *See* migration

Padma River, 62, 161
Paikgachha, 104, 116, 119, 122–23, *123,* 125, 132, 135, 137, 146, 150, 153, 186, 208n3
Pakistan Academy for Rural Development (PARD), 59–60, 211n8
Parkinson, J. R., 58
Partition, 31, 38, 133, 209n15, 214n14
patitabad, 209n13
peasantry: Bangladeshi understanding of, 156, 215n21; depeasantization, 122, 139–44, 215n17; Khulna district, peasant social movements in, 3–4, 14; *raiyat* laborers, 37, 45
Permanent Settlement area, 26, 37
Pew Research Center, 57
Piddington, Henry, 35
Polder 22: Blue Gold program in, 181–83; Canal Movement or Khal Andolon in, 175, 180; DDP (Delta Development Program) in, 50; map, *xv;* marginal laborers, most villagers as, 124; methodology of research on, 201; resistance to shrimp aquaculture in, 169–75, *172, 173, 176,* 184, 188. *See also specific inhabitants by name*
Polder 23: map, *xv;* methodology of research on, 201, 202; migration from, 122, 150–57; rice agriculture in, 213n1; shrimp aquaculture in, 125, 162, 165. *See also* Kolanihat
Polder 25, 104
Polder 29: Blue Gold program in, 181; map, *xv;* methodology of research on, 201; visions of agrarian change in, 188; zoning and, 115. *See also* Tilokpur
polder system, 39–42, 49, 87, 104, 205, 214n4, 214n9. *See also* Coastal Embankment Project
population control, 60, 211n8
Pradox (pralidoxime chloride), 146–47, *148*
"premature land reclamation," 37, 45, 210n20

Quddus, A. H. G., 149

Radhika (inhabitant of Kolanihat), 133–35, 137, 139, 214n14
Rahman, Sheikh Mujibur, 57
Rahman, Ziaur, 57
railway network, 30–31, 43
Rainey, H. James, 33–34, 209–10n16
raiyat laborers, 37, 45
Rampal Power Plant, 15–16
reclamation of land. *See* embankment and land reclamation
religion: Hindus, 124, 127, 153, 214n14, 215n21; migration for religious reasons, 124, 150, 214n14; Muslims, 48, 150, 208n4, 209n15, 215n21
Rennell, James, and *Bengal Atlas,* 23, 30, 31, 208n2
resettlement/retreat. *See* migration
resilience, 75, 100, 105, 156
resistance to embankment systems, 41, 210n23
resistance to shrimp aquaculture, 21, 158–89, 191–92; Canal Movement or Khal Andolon, 167, 176–80; harassment of resisters, 166–67, 181–82; incremental versus transformational

resistance to shrimp aquaculture (*continued*)
adaptation, 184–89; in Kolanihat, 1–3; local contestation of landless collectives, 180–81; mobilization beyond shrimp, 167–69; by Nijera Kori, 3–4, 15, 50, 141, 158, 165–66, 199, 201–2; ongoing mobilization for, 180–83; in Polder 22, 169–75, *172, 173, 176*. See also Polder 22; Tilokpur
retreat/resettlement. *See* migration
rice agriculture: *aman* rice crop, 40, 124, 138, 152, 205; attitudes of farmers versus researchers toward, 65–66, 74–76, 118–22, 156–57; bamboo rice silos, 119, *120*; collective rice farming, *173*; in development discourse, 98, 101, 110–12; *dhan* and *bhat*, 121; embankment and reclamation of land for, 35, 36; flooding and, 40; high-yield foreign rice varieties, 39; in Kolanihat, 123–24, 135, 137–39, *138*; labor and employment, 140–42, 142*t*, 164; land tenure and, 133–39, *138*; multiple crops of, 213n1; in Polder 22, *173*, 174; resistance to shrimp aquaculture and, 3; seeds and seed varieties, 175, *176, 177*, 216n10, 216n12; shrimp aquaculture versus, 1–2, 19, 21, 50, 56, 69, 71–72, 98, 101, 110–12, 116–17, 121–22, 125–30, 140–44, 142*t*, 143*t*, 164; Tilokpur, revival of rice cultivation in, 158–61, *159, 160, 163*, 164; TRM/TRM++ and, 106; zoning and, 116–17. *See also* indigenous seeds and seed varieties
rickshaw-van pulling, 164–65, 216n7
Right to Information (RTI) Act (2009), 168, 182
rights-based approach to staying in one's home, disavowals of, 10, 75
river bank erosion and sedimentation, 212n2
river level rise, 84
river systems. *See* water management and river systems; *specific rivers*
Rohingya, 210n26
Rose, Rezanur Rahman, 169, 201

Saiful (*gher* manager), 125
Saline Water Resistance Committee, 165–66
salinity of soil. *See* soil salinity, rise of
salinization/depletion of groundwater, 83, 131–32, 164
saltwater shrimp aquaculture. *See* shrimp aquaculture
scientific knowledge. *See* knowledge production and politics of uncertainty
Scramble for Africa, 69

sea level rise, 25, 78, 82, 83, 84
sedimentation of rivers, 212n2
seeds, seed varieties, and seed banks, 175, *176, 177*, 216n10
shak diye mach dhaka ("covering up the fish with greens"), 17
Shanin, Teodor, 215n21
Shatkira district, 186
Shearer, Christine, 13
shipping routes, 29–30
Shonjoy (New Town inhabitant), 152–55
shrimp aquaculture, 21, 118–57; algae skimming, 144–47, *146–48*; attitudes of workers, landholders, and researchers toward, 118–22, 156; causality and responsibility for social and ecological damage of, 213n5; deforestation and river bank erosion/sedimentation, 212n2; depeasantization due to, 139–40; in development discourse, 98, 101, 110–12; diseases threatening, *123*, 126, 136–37, 139, 142, 145, 147, 153; dispossession and, 1–4, 49–50, 71–72, 89–90, 111–12, 122, 126–30; foreign aid/development intervention encouraging, 49–50; growth of, 71; international market and prices, 102, 118, 136, 214–15n15; job loss due to, 140–44, 142*t*, 143*t*; labor and employment in, 141–49, 142*t*, 143*t*, 165; land tenure and, 133–39, *138*; landscape changes due to, 119, 130–33; methodology of research in, 201–3; migration caused by, 71–72, 119, 120–21, 122, 149–57; NGOs and, 101, 103, 122–23, *123*, 143–44, 156, 213n5; Paikgachha, shrimp culture research station in, 122–23, *123*; photo of shrimp, *13*; primacy/absence of climate change narrative from discussion of, 55–56; processing for export, 141; rice agriculture versus, 1–2, 19, 21, 50, 56, 69, 71–72, 98, 101, 110–12, 116–17, 121–22, 125–30, 140–44, 142*t*, 143*t*, 163; scientific understanding of, 89–90; soil salinity, rise of, 3, 18, 72, 78–79, 83, 116, 212n16; TRM/TRM++ and, 106–7; varieties of shrimp, 214–15n15; waterlogging and, 100–110; wild shrimp fry collection, 148–49; zoning and, 115–17. *See also* Kolanihat; resistance to shrimp aquaculture
Sivaramakrishnan, K., 208n3, 209n12, 209n13
Small, Christopher, 93–94
Smith, Zadie, 190
Smythies, E. A., 45–46
Sobhan, Rehman, 57
social reproduction, 159–60

Society of Arts, London, 30
soil erosion and accretion, 46, 84, 212n2, 212n4
soil fertility, 34–35, 214n3
soil salinity, rise of: causes of, 3, 18; drinking water supply and, 131; flushing, 83, 164, 216n5; scientific understanding of, 82–83; shrimp aquaculture and, 3, 18, 72, 78–79, 83, 116, 131, 212n16; trees and crops killed off by, 131, 132, 162–63, 166
Special Economic Zone in Mongla, 75
Stakeholder Workshops, 179–80, 200, 216n11
storm surges, 85, 93–94
Stott, Clare, 93
subinfeudination, 47–48
subsidence, 25, 37, 84, 209n7, 210n21
subsistence production, agricultural, 131, 133, 140, 174
suicide attempts using chemical fertilizers, 146–47
Sundarban Shrimp Private Limited, 89
Sundarbans: forest degradation in, 86, 90–91; Khulna district, relationship to, 208n1; map, xv; Rampal Power Plant protests and, 15–16; terminology, etymology, and usage, 208n1; wastelands, designation as, 26–27, 28. See also history of climate dystopia in Khulna
Sundarbans Act (1905), 209n9

Tala subdistrict, 104, 107–10
Temple, Sir Richard, 43–45
Temporarily Settled Estates, 36, 211n27
temporariness and impermanence of landscape: dispossession and, 43; in history of climate dystopia in Khulna, 32–33, 43; physical geography of region and, 24–26
Thailand, shrimp aquaculture in, 136
Thomas, John W., 40, 210n23
Thompson, E. P., *Whigs and Hunters*, 6, 98
tigers, in Sundarbans, 48
tilapia cultivation, 133
Tilokpur, 158–69; Canal Movement or Khal Andolon, 167, 180; growing seasons in, 216n7; harassment of anti-shrimp movement members, 166–67; historical background, 161; migration from, during shrimp aquaculture period, 158, 159, 160–61; mobilization beyond shrimp in, 167–69; movement against shrimp aquaculture in, 165–67, *167*, 184; revival of rice cultivation in, 158–61, *159, 160, 163,* 164; shrimp aquaculture in, 162–65, *163,* 215n1. *See also specific inhabitants by name*

Titash (inhabitant of Tilokpur), 158–60
transformational versus incremental adaptation, 184–89
TRM and TRM++, 104–10
Tsing, Anna, 195

UN Women, 185
uncertainty, knowledge production and politics of. *See* circulations of knowledge and uncertainty in development practice; knowledge production and politics of uncertainty
unemployment. *See* labor and employment
United Nations Champions of the Earth award, 14
United Nations Climate Change Conferences, 16, 52
United Nations Development Programme (UNDP), 104–10
United Nations Framework Convention on Climate Change (UNFCCC), 14, 74. *See also* Development and Climate Days workshop, UN Climate Change Conference
urbanization as adaptation strategy, 12–13, 53, 75, 187. *See also* migration
US National Flood Insurance Program, 213n7
USAID, 37–39, 49, 55, 59, 60, 62, 67, 75, 92, 111, 116, 122, *123,* 179, 210n22

Vietnam, shrimp aquaculture in, 136

Wajed Ali, 128, 171, 214n7
Wakil Saheb (*gher* businessman) and Wakil's *gher, 125,* 127, 129–30, 133, 134, 136, 137, 138, 140
wastelands, construction and reclamation of land as: concept of waste, 209n10; current practice of, 209n11; dispossession due to, 44, 47–48; historical practice of, 26–28, *28, 29,* 44, 47–48, 209nn11–13
water management and river systems: avulsion, 24–25, 209n6; *char* formation, 28, *28–29*; forest resources, tidal channels as means of exporting, 45–46; maintenance of, 35, 41–42, 178–79, 210n24; physical geography of region and, 24–26; shipping routes, historical development of, 29–30; shrimp aquaculture and, 129–30; silting due to embankment and land reclamation, 37, 46. *See also* canals; indigenous water management systems
Water Resources Planning Organization (WARPO), 88, 113
water supply for drinking, 131–32

waterlogging: canals and, 101, 180; defined, 100; in development discourse, 98, 100–110; mapping exercise, 100–101, 102–4; river sedimentation and, 37; scientific knowledge about, 83–84, 101; TRM and TRM++, 104–10
Watts, Michael, 91, 196
Westland, J., 209–10n16
wetland versus dryland approach, 40–41
wetlands, 40, 106, 132, 210n19, 213n3
white spot syndrome, 136
Whyte, Kyle, 207n6
"winners and losers" discourse, 188
women: as algae skimmers, 144–47, 146–48; contraceptive technologies and, 60, 211n8; in garment industry in Bangladesh, 215n20; gendered division of labor, 3, 141, 159, 215n2; labor and employment for, 141, 144–49; microcredit industry aimed at, 61; shrimp aquaculture and violence against, 126; wild shrimp fry collection by, 149. *See also* social reproduction
World Bank, 14, 57, 58, 60, 69, 70, 71, 87, 123, 207n2, 212n7
WorldFish, 68, 113, 123, 156, 211n28, 214n15

Years of Living Dangerously (TV show), 63–64
Yunus, Mohammad, 61

zamindars, 26, 27, 38, 44–47, 205, 214n4, 214n13, 215n21
Zeiderman, Austin, 207n3
zoning, 68, 98, 112–17

CPSIA information can be obtained
at www.ICGtesting.com
Printed in the USA
LVHW091905240921
698690LV00001B/6

9 781501 759154